58 Springer Series in Solid-State Sciences

Edited by Manuel Cardona

Springer Series in Solid-State Sciences

Editors: M. Cardona P. Fulde H.-J. Queisser

40 **Semiconductor Physics**
An Introduction By K. Seeger

41 **The LMTO Method**
Muffin-Tin Orbitals and Electronic Structure
By H. L. Skriver

42 **Crystal Optics with Spatial Dispersion, and Excitons**
By V. M. Agranovich and V. L. Ginzburg

43 **Resonant Nonlinear Interactions of Light with Matter**
By V. S. Butylkin, A. E. Kaplan, Yu. G. Khronopulo, and E. I. Yakubovich

44 **Elastic Media with Microstructure II**
Three-Dimensional Models
By I. A. Kunin

45 **Electronic Properties of Doped Semiconductors**
By B. I. Shklovskii and A. L. Efros

46 **Topological Disorder in Condensed Matter**
Editors: F. Yonezawa and T. Ninomiya

47 **Statics and Dynamics of Nonlinear Systems**
Editors: G. Benedek, H. Bilz, and R. Zeyher

48 **Magnetic Phase Transitions**
Editors: M. Ausloos and R. J. Elliott

49 **Organic Molecular Aggregates,** Electronic Excitation and Interaction Processes
Editors: P. Reineker, H. Haken, and H. C. Wolf

50 **Multiple Diffraction of X-Rays in Crystals**
By Shih-Lin Chang

51 **Phonon Scattering in Condensed Matter**
Editors: W. Eisenmenger, K. Laßmann, and S. Döttinger

52 **Superconductivity in Magnetic and Exotic Materials**
Editors: T. Matsubara and A. Kotani

53 **Two-Dimensional Systems, Heterostructures, and Superlattices**
Editors: G. Bauer, F. Kuchar, and H. Heinrich

54 **Magnetic Excitations and Fluctuations**
Editors: S. Lovesey, U. Balucani, F. Borsa, and V. Tognetti

55 **The Theory of Magnetism II**
Thermodynamics and Statistical Mechanics
By D. C. Mattis

56 **Spin Fluctuations in Itinerant Electron Magnetism** By T. Moriya

57 **Polycrystalline Semiconductors,**
Physical Properties and Applications
Editor: G. Harbeke

58 **The Recursion Method and Its Applications**
Editors: D. Pettifor and D. Weaire

59 **Dynamical Processes and Ordering on Solid Surfaces**
Editors: A. Yoshimori and M. Tsukada

Volume 1–39 are listed on the back inside cover

Springer Series in Solid-State Sciences

The Recursion Method and Its Applications

Proceedings of the Conference,
Imperial College, London, England
September 13–14, 1984

Editors:
D. G. Pettifor and D. L. Weaire

With 42 Figures

Springer-Verlag
Berlin Heidelberg New York Tokyo

Dr. David G. Pettifor

Imperial College of Science and Technology, Department of Mathematics,
Huxley Building, Queen's Gate,
London SW7 2BZ, England

Professor Denis L. Weaire

Department of Pure and Applied Physics, Trinity College,
Dublin 2, Ireland

Series Editors:

Professor Dr. Manuel Cardona
Professor Dr. Peter Fulde
Professor Dr. Hans-Joachim Queisser

Max-Planck-Institut für Festkörperforschung, Heisenbergstrasse 1
D-7000 Stuttgart 80, Fed. Rep. of Germany

ISBN 3-540-15173-7 Springer-Verlag Berlin Heidelberg New York Tokyo
ISBN 0-387-15173-7 Springer-Verlag New York Heidelberg Berlin Tokyo

Offset printing: Beltz Offsetdruck, 6944 Hemsbach/Bergstr. Bookbinding: J. Schäffer OHG, 6718 Grünstadt
2153/3130-543210

Preface

This volume reviews recent advances in the development and application of the recursion method in computational solid state physics and elsewhere. It comprises the invited papers which were presented at a two-day conference at Imperial College, London during September 1984.

The recursion method is based on the Lanczos algorithm for the tridiagonalisation of matrices, but it is much more than a straightforward numerical technique. It is widely regarded as the most elegant framework for a variety of calculations into which one may incorporate physical insights and a number of technical devices. The standard reference is Volume 35 of Solid State Physics, which contains all the early ideas of Heine, Haydock and others, upon which the method was established. The present volume provides the first review of subsequent developments. It also indicates where problems remain, or opinions differ, in the interpretation of the mathematical details or choice of practical techniques in applications. The field is still very lively and much remains to be done, as the summary chapter clearly demonstrates.

We are grateful to the S.E.R.C.'s Collaborative Computational Project No. 9 on the electronic structure of solids and the Institute of Physics's Solid State Sub-committee for their sponsorship of the conference. We thank Angus MacKinnon for his help in conference organisation and Jacyntha Crawley for secretarial assistance.

December 1984 *David G. Pettifor* *Denis L. Weaire*

Contents

Part V Lanczos Method Applications

Part VI Conference Summary

Part I

Introduction

Why Recur?

Volker Heine

Cavendish Laboratory, Madingley Road, Cambridge CB3 0HE, United Kingdom

The recursion method is having an ever widening development and range of applications, and in opening this conference I want to indicate what I think are some of the ideas behind this. I need hardly explain to this audience what the recursion method is, beyond defining my notation. Let me write the recursion relation as (see p.70: references purely in the form of page numbers refer to the review articles [1], [2], [3])

$$b_{n+1}\, u_{n+1} = Hu_n - a_n u_n - b_n u_{n-1} \tag{1}$$

where H is the Hamiltonian or other operator in some matrix form, the u_n constitute the set of new basis functions in column vector form, and the a_n, b_n are the coefficients. The coefficients are used to write a continued fraction to represent the matrix element of the Greenian or resolvent

$$G_{00}(E) = \langle u_0 | (E-H)^{-1} | u_0 \rangle. \tag{2}$$

Some historical remarks will illustrate the main points I want to make. The first is the realisation that in large systems like solids, formally infinitely large, the eigenfunctions of the Hamiltonian are rather useless, although most standard quantum mechanics is written in terms of them. If we consider a perfect periodic solid with a large number (of order 10^{23}) atoms, and destroy the perfect periodicity by perturbing just one atom, then this is sufficient to mess up the precise eigenstates completely. But of course the physics has not changed much. I learnt this already in the 1950's from Friedel, who noted that the physics (such as charge densities and magnetic moments) is contained in the local density of states $n(E,\underline{r})$ (pp.1-5). Moreover Friedel also pointed out that this quantity is stable under small changes of the system (pp.5, 9-17). In the 1960's one learnt that the local density of states is related to the Green function which in our basis set becomes the quantity (2) (pp.6-8). It therefore became desirable to calculate or approximate the Green function directly without first calculating the eigenfunctions. While I had already posed the question in this way (pp.123-127), it was entirely Haydock's ideas that provided the answer and led to the recursion method. He started from the work of Friedel and F. Cyrot-Lackmann who were at that time fitting Gaussian-type densities of states to a small number of calculated moments. The first applications were to calculate the local densities of states on atoms near surfaces of transition metals and in amorphous semiconductors (pp.320-356).

By integrating the local density of states up to the Fermi level one can calculate the total weight on one atom, and the magnetic moment in spin polarised systems. A current application is the calculation of the coupling J_{ij} between two atoms i and j in a magnetic metal such as iron [4], while the surrounding atoms may be held in any desired magnetic configuration. The mobile electrons washing over atom j carry with them a memory of the direction of the magnetic moment $\underline{m}_j$ and produce on atom i a small component of the magnetic moment there parallel to $\underline{m}_j$. This small component on atom i may be evaluated by rotating the moment $\underline{m}_j$ to different directions which gives directly the coupling J_{ij}. The latter is found to depend quite strongly on the magnetic configuration of the surrounding atoms, which makes it important to use the recursion method, where no special assumption needs to be made about the environment.

These simple examples already illustrate the three basic principles that make the recursion method so useful in an ever expanding range of applications. The first is that the structure of the method already incorporates as much of the right physics as possible, namely, one wants to calculate directly the quantity (2) which is stable under small changes in the system rather than unstable as the eigenstates are. Secondly, in a large system such as all solid state physics, one is always overwhelmed by too much information, in principle an infinite amount. The trouble with computers is that they give too many numbers, whereas physically one wants some combined quantity such as a magnetic moment. One does not want all the eigenstates, one does not even want thousands of eigenvalues. Although one uses clusters with N equal to a thousand or more atoms, one only recurses about 20 times to obtain a very good representation of the local density of states. A complete calculation of the eigenvalues would require N recursions! Thirdly, one has an unlimited freedom to choose the starting vector u_0 to suit one's problem. In the above examples it was always an atomic orbital with suitable spin direction, but much more general starting vectors can be used.

The third point is illustrated by the calculation of photoemission from transition metal d-bands into one assumed plane wave state in a higher band. This is not a local atomic phenomenon, and contains selection rule and matrix element effects depending on the final state and on the five d-orbitals. Haydock showed that all these effects can be incorporated into a specific starting vector u_0 (pp.86, 377). This is currently being applied to photoemission from magnetic metals above and below the Curie temperature [5].

We gain a somewhat different perspective on the usefulness of the recursion method by considering the new basis set u_0, u_1, u_2, ... in the light of the first and third principles. Of course we are not limited to quantum mechanics or atomic orbitals; anything which can be formulated as a linear eigenvalue problem will do, but it is convenient to use the language of quantum mechanics. We have already said (principle 3) that u_0 is chosen to correspond to the physical quantity desired. The set of u_n is then in a certain sense the <u>best</u> basis set for calculating the required G_{00} (E) for the following reason. The u_n are the functions Hu_0, H^2u_0, H^3u_0 etc., each orthogonalised to the previous ones. The extra factor of H each time means that the u_n are successively more remote from u_0 in the function space of H, and they therefore form an optimised basis set for calculating the quantity (2). That is why a relatively small number of recursions suffices to give well converged answers. The idea that the u_n are successively more disconnected from u_0 as far as H is concerned may be a somewhat abstract concept, but it can be illustrated (purely illustrated) by considering u_0 as an initial wave function at t = 0 and asking for its time development $\psi(t)$. The latter can be written as a power series in t by expanding the time development operator $\exp(-iHt/\hbar)$, the n'th term being

$$t^n (i\hbar)^{-n} H^n u_0/n! \qquad (3)$$

Thus $H^n u_0$, which is equivalent to u_n, only affects the n'th derivative of $\psi(t)$. Another way to visualise the increasing remoteness of the u_n is to consider a solid with a single s state on each atom with nearest neighbour hopping. The operation of H means that each u_n reaches out to one further shell of neighbours, and the orthogonalisation means there is rather a hole in the middle (or rather a mess which is very orthogonal to u_0 and anything closely connected with it) (p.71 and discussed by O'REILLY at this conference).

The optimal property of the basis set u_n is especially important in many-body and N-body problems. Consider for example the antiferromagnetic linear chain of spin ½ atoms. There are 2^N basis functions which quickly become unmanageable as N increases if one considers all of them. However, one need not. Two simple basis functions can be represented diagrammatically thus:

$$\ldots \uparrow \downarrow \uparrow \downarrow \uparrow \downarrow \uparrow \downarrow \ldots \qquad (4a)$$

$$\ldots \uparrow \downarrow \uparrow \underline{\uparrow \downarrow} \downarrow \uparrow \downarrow \ldots \qquad (4b)$$

Here (4a) is the zero order antiferromagnetic state and (4b) is the same with one pair of spins switched. The H operating on (4a) only introduces switches like (4b). Thus starting with (4a) as u_0, the u_1 is a sum of states like (4b) each with one switch. Of course there are N states with one switch each, but u_1 is a single state which is the correct linear combination for coupling to u_0: the rest are irrelevant. Similarly, u_n contains up to n switches. Now the correct ground state is known to contain about 5% of atoms switched. (Paradoxically, this means that each term in the expansion of the ground state looks very like the simple (4a) but the number of terms is so large that the actual weight of (4a) in the ground state is effectively zero.) Thus to ensure a good representation of the ground state wave function one should go somewhat beyond 0.05N switches, say to 0.1N recursions. This is vastly less than the 2^N recursions needed for an exact diagonalisation of H [6].

I want to make a few remarks about the relation of the recursion method to the Lanczos algorithm for diagonalising a matrix. Both use the same algorithm (1) and the dividing line between the two has become a bit blurred over the years, but there are two points at issue. The Lanczos method in its original form involved finding <u>all</u> the eigenvalues and hence recursing N times to 'completion' for a matrix of order N. In the recursion method the starting vector u_0 plays a key role, often in the sense of defining the matrix element (2) but in any case in the sense of creating the new basis set of functions u_n which become increasingly remote from the physics being described. It is this latter property which means one only requires a moderate number of recursions, much less than N. This is for me the distinguishing feature of the recursion method.

Incidentally the recursion method can also be used for finding eigenvalues. Suppose we have a matrix of N = 2000 and require the lowest 40 eigenvalues. I am assuming that the matrix is well-conditioned in the sense that one can identify say a 60 × 60 submatrix which already contains a good approximation to the lowest 40 eigenvectors. We diagonalise the 60 × 60 exactly and define u_0 as the sum of the first 40 of these eigenvectors. This ensures that the final required eigenvectors have a substantial projection onto u_0 and hence that they converge satisfactorily. One then iterates about 120 times to generate a continued fraction of length 120 which can be rationalised to 120 simple poles. Of these, the lowest 40 will correspond well to the lowest 40 eigenvalues of the full matrix,while the remaining 80 suffice to represent the rather low projected density of states (projected onto u_0) over the rest of the eigenvalue spectrum. The 40 full eigenvectors can also be recovered from the set u_n. This kind of technique has recently been used by R. MARTIN, R. NEEDS, R. HAYDOCK and C. NEX [7] for a total energy calculation on metallic and molecular hydrogen using a large basis set of plane waves. It is similar to what has long been done by WHITEHEAD [8] in nuclear physics. The point is that the computing time increases only as N^2 as one increases the size N of the basis set (for a constant number of eigenvalues), whereas a full diagonalisation requires a time of order N^3.

I have talked about the recursion method in the context of very large matrices: one of its important features is that at the opposite extreme one can also do purely analytic calculations and in fact cover all the range in between. Two aspects of the recursion method make this possible. Firstly one can carry out the first one or two (or perhaps three) steps of the recursion analytically, and the form of the continued fraction relates directly to a diagrammatic development of the particular matrix element (2) of the resolvent operator. This is not emphasised very much in the set of review articles already referred to, but can be found better in the original paper by HAYDOCK, HEINE and KELLY [9]. Another example is the work of KELLY [10] (also pp.333-5). Secondly one has the 'square root terminator' (p.314) and other terminators (see this conference) which one can apply at any desired low order of the continued fraction as a useful approximation for the remainder of the fraction to infinity. An example being considered by FOULKES [11] is shown in the figure with a central atom bonding to two blocks of solid on the left and right. It is supposed to represent an impurity segregated in a grain boundary of a metal. If for simplicity we take only a single orbital on the central atom chosen as u_0, we can write down exactly

$$G_{00}(E) = [E - h_1^2\, G_{11}(E) - h_2^2\, G_{22}(E)]^{-1} \qquad (5)$$

where h_1, h_2 are the matrix elements coupling u_0 to the surface orbitals in the sense of KELLY [10] (pp.333-5) on the left and

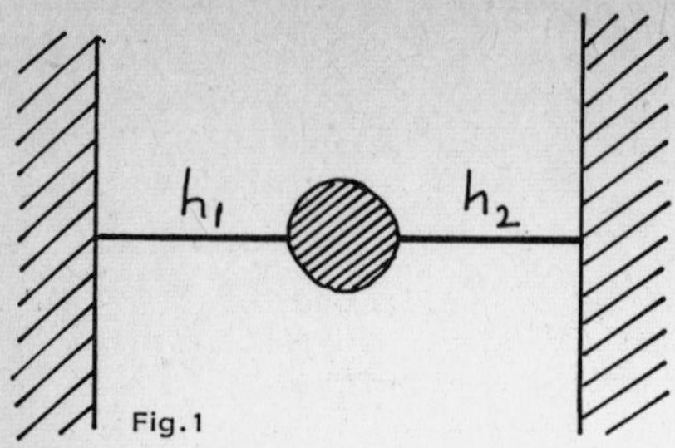

Fig.1

right. Moreover G_{11} and G_{22} are exactly the Green functions for the left and right blocks for the respective surface orbitals. One can also calculate analytically the change in total energy for cutting the h_1, h_2 bonds, i.e. the bonding energy between the two grains due to the presence of the impurity atom, using the techniques of pp.102-5, 277-81, 336-9 of the review articles.

Finally I would like to look into the future: the Lanczos or recursion method may become the fastest way of diagonalising very large matrices [8]. Although further Very Large Scale Integration may increase computer speed by an order of magnitude, there may be a limit to the amount of extra computer power that can be squeezed out that way. The significant wave of the future may be parallel processing using dedicated processors for specific jobs. Instead of instructing the computer operator to load your own tape, you may be telling him to plug in your chip. The recursion algorithm is particularly appropriate for this. The basic step is the matrix multiplication Hu. Let us imagine a processor for handling a matrix of order N = 1000. The n'th column of N elements of H are loaded as one string from disc onto N registers, multiplied simultaneously by the n'th element of u, and added simultaneously into what has already been accumulated of the product vector Hu. With floating point multipliers costing about 100 dollars each, it is not inconceivable to have 1000 of them in parallel. The point about the Lanczos algorithm is that the multiplication of one row of H is totally independent of any other row so that they can go on truly in parallel.

In conclusion, the recursion method is still undergoing significant development and finding wider forms of application, as the programme of this conference witnesses. I hope I have caught some of the flavour of this in these brief remarks. I apologise to all those whose interests I have not mentioned, or whose work I have not specifically referenced: this is not supposed to be a survey. Indeed, the conference proceedings as a whole will update the reviews [1], [2], [3], and I want to express my gratitude on behalf of all the participants to the organisers: Roger Haydock, Denis Weaire, and particularly Angus MacKinnon and David Pettifor who have done most of the work.

1. V. Heine: Solid State Physics 35, 1 (1980).

2. R. Haydock: Solid State Physics 35, 216 (1980).

3. M.J. Kelly: Solid State Physics 35, 296 (1980).

4. L.M. Small and V. Heine: J.Phys.F, to appear (1984 or 1985).

5. E.M. Haines, V. Heine and A. Ziegler: J.Phys.F, to appear (1985).

6. R. Haydock and C. Nex, unpublished.

7. R. Martin, R. Needs, R. Haydock and C. Nex: to be published.

8. R.R. Whitehead: this conference.

9. R. Haydock, V. Heine and M.J. Kelly: J.Phys.C 5, 2845 (1972).

10. M.J. Kelly: J.Phys.C 7, L157 (1974); Surf.Sci. 43, 587 (1974).

11. M. Foulkes: to be published.

The Recursive Solution of Schroedinger's Equation[1]

Roger Haydock[2]

Institute of Theoretical Science, University of Oregon, Eugene, OR 97403, USA

1 Strong Coupling

The recursion method [1] gives a convergent sequence of bounded approximants to the solution of Schroedinger's equation and other linear, hermitian equations. This is the best alternative to an exact solution in closed form. The purpose of this paper is to describe the systems where such a method is necessary, review the method, and discuss some applications.

To explain when such a method is necessary, let us consider several quantum mechanical systems: electrons in atoms, molecules, and solids, keeping in mind the aspects of these systems which are of a general nature. Starting with atoms, the one-electron case is exactly solvable, and in the many electron atom, if we make the mean field approximation we obtain potentials which are similar to the one electron atom in that, for each angular momentum, levels are only shifted by a small amount relative to their spacing. This is an example of a weak-coupling problem, and perturbation theory is an appropriate method of solution.

Going from the electronic structure of atoms to that of molecules, we confront the problem of interactions between atomic levels, which are strong relative to their spacing. This comes about in two ways: first from interaction between similar levels on different atoms, and second from the splitting of atomic angular momentum multiplets by breaking of spherical symmetry. Although we can resolve the molecular orbitals into different representations of the molecular symmetry, there are no exact solutions for molecular potentials from which to perturb. Thus,this problem belongs to the class whose characteristic is strong coupling between a finite number of levels. This kind of problem must be solved variationally, with functions which span the strongly coupled degrees of freedom. Fortunately in this case there are only a finite

---

1 Supported in part by the National Science Foundation(USA) Condensed Matter Theory Grant DMR 81 22004.

2 SERC Senior Visiting Fellow, Cavendish Laboratory, Madingley Road, Cambridge, CB3 0HE, UK.

number of these degrees and, using a linear basis, the variational part of the problem reduces to diagonalizing a finite matrix, which can be done numerically to arbitrary accuracy. Once the strong coupling has been diagonalized, the problem can be solved by perturbation theory.

Consistent with the increasing complications of atoms and molecules, solids present the problem of strong coupling of 10^{23} or an infinite number of degrees of freedom. For perfectly crystalline solids, symmetry can be used to reduce the problem to the finite strong coupling problem within a single unit cell, and thus to diagonalization of a finite matrix. However, in the presence of disorder or any defect, such as a surface, which breaks translation symmetry, the problem is once again of the infinite strong coupling class. In terms of a linear variational basis, the problem is that of diagonalizing an infinite matrix.

It is in the large or infinite strong coupling case that the recursion method is necessary. Perturbation theory does not converge and thus can produce any answer. Cluster methods, in which a finite sub-matrix is used as an approximation, depend on the intuition of the person choosing the sub-matrix. Non-linear variational methods also depend on intuition, and functional integral techniques lead to uncontrolled approximations.

It is in the infinite strong coupling case that new physics arises, because it is in this case that there are true phase transitions, abrupt changes in the physical properties of the system. In weak coupling, properties can only change slightly and only in a smooth way. In finite strong coupling, states can change their symmetry, but they must do so smoothly unless there is an exact symmetry of the hamiltonian which allows them to cross without interacting. Thus in condensed systems where we want to understand the relation between various ground states and their excitations, the recursion method is the appropriate way to proceed. Unlike other approaches to infinite strong coupling, it has a firm mathematical basis and provides bounds for approximations to many quantities.

For simplicity the illustrations of weak and strong coupling at the beginning of this section used the mean field approximation. However, the same classification applies to the full manybody problem, except that it is harder to find cases of weak coupling. Correlation between electrons is greatest where there are many configurations with nearly the same energy, the strong coupling case. However, for atoms and molecules the strong coupling is still finite,though involving more degrees of freedom than in the mean field approximation.

Infinite strong coupling for the manybody problem still only arises in the solid.

Having described in a general way what sort of problems require the recursion method, the rest of this paper reviews the method and relates it to other methods. The time independent Schroedinger equation and its solution by means of the resolvent are introduced in section 2. In section 3, we discuss the starting state, what properties it must possess, and how it relates to the physics under investigation. Section 4 contains a description of the method and its relation to the mathematical moment problem. The computational aspects of the method, termination and the Lanczos method are covered in section 5. Section 6 surveys the various physical quantities which can be calculated from recursion, and in the last section, 7, we mention some trends in development of the method and what problems may now be solved.

2 Time Independent Schroedinger Equation and the Resolvent

The recursion method solves the Schroedinger equation in its time independent form, that is in terms of the stationary states of the system. For systems with time independent hamiltonians, time dependence can be recovered by a fourier transform for which the recursion method provides a convenient quadrature. Time dependent hamiltonians can be converted to time independent ones by including the driving mechanism in a hamiltonian which conserves energy as a whole, although energy is transferred back and forth between two sub-systems.

Thus quantum problems may be expressed in the equation

$$H \psi = E \psi \tag{2.1}$$

where the stationary states $\{\psi\}$ depend on various coordinates of particles and have the appropriate symmetry, it is the hamiltonian which specifies the laws of motion, and $\{E\}$ are energies of stationary states. If ψ contains only a finite number of degrees of freedom, then 2.1 can be written as an eigenvalue equation for a finite matrix. However, if ψ has an infinite number of degrees of freedom as in the infinite strong coupling problem, then in general this equation is awkward to use and it is better to write an equation for the resolvent, which is equivalent to 2.1 in the finite dimensional case. By the way, the number of degrees of freedom in ψ must always be countable or the problem is ambiguous.

Calculation of the resolvent or greenian (fourier transform of the green function in operator form) is a way of posing the quantum problem which applies to both finite and infinite systems. The resolvent operator is,

$$R(E) = (E - H)^{-1} , \qquad (2.2)$$

where E is a complex energy variable. This operator is an analytic function of complex E, with singularities in some regions of the real E-axis. It is equivalent to 2.1 in that it completely describes the motion of the system.

$$\left|(2\pi i)^{-1} \int v^{\dagger} R(E)\, u\, e^{-iEt/h}\, dE\right|^2 \qquad (2.3)$$

is the probability that the system, initially in state u, is found in state v after time t, where the integral is over a contour in the complex E-plane which encloses all the singularities of R(E), and where

$$x^{\dagger} y = \int dr\, x(r)^{*}\, y(r) , \qquad (2.4)$$

is the inner product between states.

In an infinite system R(E) contains far more information than is ever wanted. Experiments involve only a few of the many possible states and it is on those we wish to concentrate in the next section.

3 The Starting State

It is important to emphasize that each quantum problem has two parts: the hamiltonian H specifies the laws of motion, and the starting state u_o specifies which kinds of motion are relevant to the problem. For example, if we want to know the electronic contribution to the potential energy of a helium atom near a surface, then u_o is the electronic ground state of the He. If we want the density of states of a crystal, then u_o must span the unit cell. If we want forces between two atoms, then u_o must span the two atoms. In essence u_o poses the specific question while H describes the model.

The choice of u_o reduces the problem from that of calculating all of R(E) to that of calculating only the state $R(E)u_o$. Those states which H or R(E) do not mix with u_o, because of symmetry or because they belong to a different phase in a manybody system, can be neglected and effort concentrated on those with which u_o communicates.

In order to apply the recursion method, u_o must also have the property that the nth moment,

$$\mu_n = u_o^{\dagger} H^n u_o , \qquad (3.1)$$

is finite for all n. The resolvent is perfectly well defined on states for which μ_n becomes infinite; the problem with these states is at some point, multiplying them by the hamiltonian, the only operation possible with a general hamiltonian, produces a state of infinite norm. Physically, the finiteness of the moments means that μ_o contains exponentially or more rapidly decreasing components of high energy states. If H has a highest energy, its moments are always finite. The problem can only arise for hamiltonians which have states up to infinite energy, as, for example, the free particle. For such systems u_o must be modified until it has finite moments. This cannot affect the physical content of u_o since it only means subtracting high energy components from u_o, but in general this is an unsolved problem, even for the hydrogen atom.

The choice of u_o defines a subspace of the states of the system, namely the space swept out by $R(E)u_o$ as E varies. This set of states may be alternatively characterized as the smallest set of stationary states needed to construct u_o. For this reason it is called the smallest invariant subspace containing u_o.

In addition to being much smaller than the full space of states, this subspace has no degeneracy. This is an immensely important point, which gives the recursion method its power. Degenerate subspaces have varying dimensionality depending on how many orthogonal stationary states can be constructed within each one. u_o has a unique projection on to each degenerate subspace, hence only that projection is necessary to construct u_o. Thus the space of states swept out by $R(E)u_o$ is non-degenerate. It is important to emphasize that this non-degeneracy is an exact property of H and u_o, not an approximation.

Lack of degeneracy is a property of semi-infinite systems with a single position coordinate and infinite degeneracy is the problem with infinite systems in more than one position coordinate. Thus on symmetry grounds the space swept out by $R(E)u_o$ is compatible with a semi-infinite model with a single position coordinate, a 1D model.

This now sets the stage for construction of the 1D model in the next section.

4 Recursion

We now construct a simple discrete 1D model for H in the subspace swept out by $R(E)u_o$, which we have shown to be non-degenerate and thus compatible with such a model. Given H and u_o which define the recursion, there is only one operation which combines them, multiplication to obtain a new vector Hu_o. There are many reasons for preferring an orthogonal basis for our model, so we orthogonalize to get,

$$x_1 = H u_o - a_o u_o , \tag{4.1}$$

where,

$$a_o = u_o^\dagger H u_o . \tag{4.2}$$

A normalized basis is also convenient, so we introduce a complex normalization b_1,

$$u_1 = x_1 / b_1 , \tag{4.3}$$

where

$$x_1^\dagger x_1 = b_1^\dagger b_1 . \tag{4.4}$$

Note that $b_1^\dagger$ is the complex conjugate of b_1, and that b_1 has an arbitrary phase, which can be chosen to make b_1 real. This is the first step of the recurrence, and has produced two matrix elements, a_o and b_1, and a new basis state, u_o.

In the general step, given u_n, u_{n-1}, and $b_n^\dagger$, we construct two more matrix elements, a_n and b_{n+1}, and a new basis state u_{n+1}. The recursion proceeds by multiplying u_n and subtracting its components on u_n and u_{n-1},

$$x_{n+1} = H u_n - a_n u_n - b_n^\dagger u_{n-1} , \tag{4.5}$$

where

$$a_n = u_n^\dagger H u_n . \tag{4.6}$$

It can be shown by the hermiticity of H that $b_n^\dagger u_{n-1}$ is the component of $H u_n$ on u_{n-1} and that $H u_n$ has no components on $u_{n-2},\ldots,u_o$. Normalizing the remainder gives,

$$u_{n+1} = x_{n+1} / b_{n+1} , \tag{4.7}$$

where b_{n+1} is a complex number satisfying,

$$x_n^\dagger x_{n+1} = b_n^\dagger b_{n+1} . \tag{4.8}$$

Once again there is an arbitrary phase in b_{n+1}, which can be chosen to make it real.

We can summarize the recursion in the hermitian three term recurrence relation,

$$0 = b_n^\dagger u_{n-1} + (a_n - H) u_n + b_{n+1} u_{n+1} , \tag{4.9}$$

with the boundary condition that $b_o^\dagger u_{-1}$ is zero. Alternatively this may be written as a matrix,

$$H_u = \begin{bmatrix} a_o & b_1^\dagger & 0 & \cdots & & & \\ b_1 & a_1 & b_2^\dagger & \cdots & & & \\ 0 & b_2 & & & & & \\ & & & b_n^\dagger & 0 & & \\ & & \cdots & b_n & a_n & b_n^\dagger & \cdots \\ & & & & 0 & b_{n+1} & \cdots \end{bmatrix} \tag{4.10}$$

where H_u is the representation in terms of $\{u_n\}$ of H restricted to the smallest invariant subspace containing u_o. Or finally, it can be expressed pictorially as shown in Fig. 1. In this way its

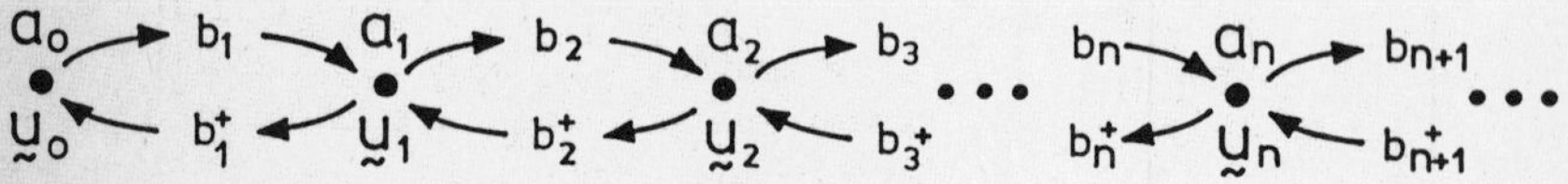

Fig. 1
Pictorial representation of the three-term recurrence.

one-dimensional character is emphasized and we see that the recursion has reduced the general hamiltonian to hopping between discrete states on a semi-infinite chain model. In this form it is possible to see intuitively how the parameters, $\{a_n\}$ and $\{b_{n+1}\}$ affect the behaviour. Large b's mean the system diffuses rapidly along the chain, while large changes in the a's are like barriers to this quantum diffusion.

The earliest known use of a three-term recurrence was by the Ancient Greeks to find the lowest common divisor of two natural numbers (positive integers). Reorganizing the arithmetic to emphasize its similarity to the recursion method, we replace H by a real number, r, between nought and one. The natural numbers $U_1<U_2<U_3<\ldots<U_n<\ldots$ replace the $\{u_n\}$ and the recursion proceeds by finding the smallest natural number, A_n, such that,

$$U_{n+1} = A_n U_n + U_{n-1} \ , \tag{4.11}$$

and rU_{n+1} is no further from a natural number than rU_n. In fact the sequence $\{rU_n\}$ comes closer and closer to being a sequence of natural numbers as n increases. If r is a rational number, then the recursion terminates when rU_n is a natural number. There is a deep relationship between the recurrence for the rational approximants to a real number and the rational approximants to a resolvent.

I learned about the three-term recurrence from the classical moment problem [2]. The moments, defined in 3.1, are the matrix elements of H between the non-orthogonal set of states

$\{H^n u_o\}$. The solution of the moment problem is essentially the construction of the same three-term recurrence as in the recursion method, but using the moments and $\{H^n u_o\}$. The difference between the recursion method and the solution to the classical moment problem is only the basis in which H is expressed.

Lanczos [3] independently arrived at the same three-term recurrence by considering the characteristic polynomial of an operator. This polynomial corresponds in the moment problem to the last element of the orthogonal basis, the one containing all the $\{H^n u_o\}$. The construction Lanczos employed corresponds exactly to that in the moment problem.

The above examples are only three from many where the three-term recurrence has been developed independently. What is surprising is that this simple idea solves so many problems.

5 Approximations

The discussion has so far been, in principal, exact. However, since much work with the recursion method is numerical, it is important to consider approximations which make finite computations possible.

The most basic approximation in numerical work is the finite precision of the real arithmetic used in computers. The absence of terms in u_{n-2},... in 4.9 depends on exact orthogonality and thereby, exact arithmetic. With finite precision arithmetic orthogonality of the $\{u_n\}$ deteriorates with the convergence in the $\{u_n\}$ of any eigenstates H may possess. That loss of orthogonality is a relatively minor problem [4] as long as the three-term recurrence, 4.9, remains accurate. This really depends on the calculation of $H u_n$ and to a lesser extent on the scalar products for a_n and b_{n+1}, for which the arithmetical precision must be sufficient.

In many applications it is also necessary to truncate H to a finite matrix. If H is sparse, then the $\{u_n\}$ will be the same for the truncated and infinite H up to some $n = N$ after which they will differ. In principle, one should stop the calculation at the Nth recursion; however, the effects of truncation seem to be small up to $n = 2N$ or $3N$, when extremal eigenvectors of the truncated H start to converge in the $\{u_n\}$.

Thus the nature of H, arithmetic precision, and available storage fix a limit on the number of recursions which can be performed. This defines the problem to which the rest of this section is devoted: how to obtain the best estimates of various quantities from a finite number of recursions and other information. Let us express this more precisely by saying that $u_o, u_1, \ldots, u_N$, $a_o, a_1, \ldots, a_N$, and $b_1, \ldots, b_{N+1}$ have been computed and that $u_{N+1}, \ldots$, $a_{N+1}, \ldots$, and $b_{N+2}, \ldots$ have not. From this we wish to estimate $R_u(E)$, the resolvent restricted to the invariant subspace containing u_o.

5.1 Quadrature

In the absence of any knowledge about H other than the computed recurrences, we apply the quadrature method [5]. Since we know nothing about the recurrence subsequent to the Nth, we must explore all possible continuations of the recurrence. It can be shown that the computed part of the recurrence determines bounds for $R_u(E)$ in the subspace spanned by $u_o, \ldots, u_N$ for E not real. Although there are no bounds for $R_u(E)$ at specific energies on the real axis, integrals of $R_u(E)$ for real E are bounded and these bounds converge exponentially with N.

As an example, the projected density of states is defined as,

$$n(E) = \pi^{-1} \lim_{\mathrm{Im}E \to 0^-} \mathrm{Im}\, u_o^\dagger R(E)\, u_o \quad . \tag{5.1}$$

The indefinite integral of this quantity,

$$N(E) = \int_{-\infty}^{E} n(E')\,dE' \tag{5.2}$$

is bounded for each value of E. Although n(E) is not bounded, the derivatives of the bounds $N^+(E)$ and $N^-(E)$ provide estimates of n(E) which are related to averages of the exact n(E) over polynomials of degree 2N.

A cruder application of the above ideas of bounds is to calculate a typical value of $R_u(E)$ by means of an arbitrary extension of the recurrence. The simplest of these is to set a_{N+1}, b_{N+2} to zero, and to evaluate the resolvent at complex energies.

5.2 Termination

Exciting new developments have taken place in the application of the recursion method to a hamiltonian, about which more is known than just its first N recurrences. This is always the case in physics, because one does not pick H out of thin air, it must be consistent with a large body of knowledge about the system under investigation. Generally, this extra information describes its asymptotic properties, the form stationary states must take in various limits such as: at large distances, or for certain values of energy corresponding to thresholds, etc.

In the first paper on the recursion method [6] the recurrence was extended with constant a's and b's corresponding to a band whose only singularities were a square-root behaviour of the density of states at the band edges. Beer and Pettifor [7] have improved the methods for estimating energies of band edges, and Turchi et al [8] have extended the method to densities of states with one or more gaps.

In the most recent work [9] we construct a model hamiltonian H^* in terms of its stationary states and their energies. A starting state u_o^* is chosen so that the invariant subspaces of the model converge to those of H and u_o for large n. In contrast to previous terminations, this one not only produces $\{a_n\}$ and $\{b_n\}$, but also $\{u_n\}$ so that, in addition to the projected density of states, the whole of $R_u(E)$ is convergently approximated. Some aspects of this are discussed in C.M.M. Nex's article in this volume.

5.3 Lanczos Method

Turning to the special case of localized states, we can place the Lanczos method in context with the recursion method. Localized states, of which eigenstates are an example, have components on the $\{u_n\}$ which decrease with n, in contrast to continuum states whose components do not decay. Because of their decay, localized states can be computed to a given accuracy in a finite number of recursions, although that number may rise rapidly with accuracy for weakly localized states. Thus, when converged, these states may be extracted from a calculation without any concern for extending the recurrence. In practice the Lanczos method is usually applied to finite matrices, for which there are only eigenstates whose components on the $\{u_n\}$ decay exponentially.

5.4 Perturbed Recurrences

Having dismissed perturbation theory for the strong coupling problem, let us now revive it in a special form for the recurrence. Suppose we consider a hamiltonian,

$$H = H_o + \lambda V \tag{5.3}$$

where λ is a dimensionless expansion parameter. Let u_o be a starting state which, in addition to the properties described in section 3, determines an invariant subspace, which varies smoothly with λ near zero. To achieve this, it is sufficient for u_o to be complete, in the sense that it contains a component from every degenerate subspace of H_o. Under these conditions the recurrence parameters and basis elements can be expanded in powers of λ either as described in [1] or in [10]. The convergence of this expansion for some finite value of λ is proved in [11], and is a consequence of the discreteness of the recurrence basis. Terminators may also be viewed as a perturbation of the recurrence.

6 Calculations

Having discussed the various approximations, we mention how these may be used in different calculations.

The projected and average densities of states have been the main purpose of recursion calculations since the beginning of the method. The density of states projected on u_o is given by 5.3. The average density of states may be computed by constructing u_o to have a weight in each degenerate subspace which is proportional to the dimension of that subspace, making the projected density of states also the average in this case.

Another kind of calculation is that of the 'bandstructure' energy density,

$$\varepsilon = \int_{E}^{E_F} n(E)\, dE \quad , \tag{6.1}$$

where n(E) is the average density of states. This has been applied to the calculation of energy differences between various crystal structures and to the frequencies of phonons. The energy of an impurity or similar local defect is a subtler quantity, being a difference in total 'bandstructure' energies. In this case

$$\Delta\varepsilon = \int_{-\infty}^{E_F} (E - E_F)\; d\{\ln[R_o(E) \,/\, R_1(E)]\} \quad , \tag{6.2}$$

where $R_o(E)$ and $R_1(E)$ are projected resolvents with and without the defect.

The third kind of calculation is that of charge density or spin density. The charge contained in an orbital, u_o, with projected density of states n(E), is

$$\rho_o = \int_{-\infty}^{E_F} n(E)\; dE \quad , \tag{6.3}$$

an expression very similar to the total energy. However, for many purposes, the entire charge density is needed, not just that projected on a single orbital. Applying the new terminator to give $\{u_n\}$, beyond those calculated exactly, gives,

$$\rho(r) = \sum_{n,m=0}^{\infty} u_n(r)\; u_m(r)^{\dagger}\; (2\pi i)^{-1} \int u_n^{\dagger}\; R(E)\; u_m\; dE \quad , \tag{6.4}$$

where the integral is over a contour which encloses the occupied states, and $u_n(r)$ is the state u_n in the position representation. This expression for $\rho(r)$ can be simplified by generating the sum from recurrence relations.

Finally, by applying the theory of perturbed recurrences, quantities related to phase-transitions may be computed. For example, if a random potential λV is added to the hamiltonian for a crystal, H_o, corrections to the recurrence can be calculated to first and second order in λ for large n. These may, in turn, be used to calculate critical energies and asymptotic forms of wavefunctions as in [10].

In the four examples above, we have progressed from the calculation for which the recursion method was devised to recent and controversial applications. The other articles in this volume suggest still further uses of the method.

7 The Future

There have been three recent developments which open up a wide range of new problems to the recursion method.

The first of these is the pioneering of dedicated computers by R.R. Whitehead and his group at Glasgow, which is described in his article in this volume. There is already at least one other group embarking on a similar project, and others will shortly follow. The recursion algorithm is suited to parallel computation, and thus, cheap units can perform vast tridiagonalizations at speeds comparable to the fastest general-purpose computers. We can expect to find more people recurring and doing bigger recursions as a result of this development.

The second advance is the development of practical terminators. This problem has been simmering since the beginning of the method, and a number of people have contributed to the solution. The basic idea of a model system whose resolvent is known analytically being unravelled to form the terminator needs investigation and further refinement. However, it opens the way to accurate work on complex spectra and a strong connection to band theory from which many terminators will come.

The third advance is the development of the theory of the perturbed recurrence. This again draws on the work of many people and is connected with the idea of terminators. We are now in a position to calculate how a perturbing potential alters the essential singularities of the spectrum. This is the central problem in the study of phase-transitions, as these essential singularities mark changes in the nature of the stationary states. As we know from the study of terminators, the asymptotic part of the recurrence determines these singularities, and so must bear some relation to a renormalization group treatment of the same singularities, but without having to make assumptions about relevant variables.

Indeed, this seems an exciting time to be working on the recursion method. A great deal of interesting physics should be revealed in the next few years.

References

1. R. Haydock: Solid State Physics 35 ed. Ehrenreich, Seitz and Turnbull (Academic, New York 1980) p.215.
2. J.A. Shohat and J.D. Tamarkin: 'The Problem of Moments' Math.Surv.I rev.ed.(Am.Math.Soc. Providence Rhode Island 1950).
3. C. Lanczos: J.Res.Natl.Bur.Stand. 45 255(1950).
4. J. Cullum and R.A. Willoughby: J.Comp.Phys. 44 329 (1981).
5. C.M.M. Nex: J.Phys.A 11 653 (1978).
6. R. Haydock, V. Heine, M.J. Kelly: J.Phys.C 5 2845 (1972)
7. N. Beer and D. Pettifor: Proc. NATO Adv.Study Inst., Gent (Plenum 1982).
8. P. Turchi, F. Ducastelle, and G. Treglia: J.Phys.C 15 2891 (1982).
9. R. Haydock and C.M.M. Nex: to be published.
10. R. Haydock: Phil.Mag.B 43 203 (1981).
11. R. Haydock: to be published.

Part II

Asymptotic Behaviour

Asymptotic Behaviour of Continued Fraction Coefficients Related to Singularities of the Weight Function

Alphonse Magnus

Institut Mathematique U.C.L., Chemin du Cyclotron 2
B-1348 Louvain-La-Neuve, Belgium

In the continued fraction expansion

$$G_0(z) = \int_\sigma (z - E)^{-1} n_0(E)\,dE = 1/(z-a_0-b_1^2/(z-a_1-\ldots)),$$

one examines relations between features of the local density of states (weight function) n_0, positive on σ, and the coefficients $\{a_n, b_n\}$.

The asymptotic description of a_n and b_n is based on two elements :

- The main asymptotic behaviour depends on the band structure of σ : $\{a_n\}$ and $\{b_n\}$ converge towards limits in the single band case, oscillate endlessly in a predictable way in the multiband case (Ducastelle).
- Damped oscillations are created by isolated singularities of n_0. The period of the oscillations in related to the position of the singularity, the rate of damping is related to the nature of the singularity.

The present study contains a description of the effects of Van Hove, algebraic (or Jacobi), and Lifshitz singularities.

1. Spectral description of a Jacobi matrix.

When the recursion method is applied to a selfadjoint operator H, with a starting state $|U_0>$, one builds the orthonormal sequence of states $|U_0>$, $|U_1>$, $|U_2>$,... with ([17], §1) $H|U_0> = a_0|U_0> + b_1|U_1>$,

$$H|U_n> = a_n|U_n> + b_{n+1}|U_{n+1}> + b_n|U_{n-1}>, n\geq 1.$$

This defines $|U_n>$ as a polynomial in H acting on $|U_0>$: $|U_n> = p_n(H)|U_0>$, $p_0(H) = I$, $p_1(H) = (H - a_0)/b_1$, etc... These polynomials depend only on the sequences $\{a_n\}$ and $\{b_n\}$, they are generated by the *recurrence relations*

(1.1) $b_1p_1(z) = z - a_0$, $b_{n+1}\, p_{n+1}(z) = (z-a_n)\, p_n(z) - b_np_{n-1}(z)$, $n \geqslant 1$.

On the Hilbert space made of square summable combinations of the states $|U_n>$, H appears as a tridiagonal operator, or Jacobi matrix

$$J = \begin{bmatrix} a_0 & b_1 & & \\ b_1 & a_1 & b_2 & \\ & \ddots & \ddots & \ddots \end{bmatrix}$$

As the recursion method is mainly used for the study of spectral properties of operators, let us consider the spectral decomposition of H ([2] chap. 4, [34] chap. 7) $f(H) = \int_\Sigma f(E)dP(E)$, where Σ is the *spectrum* of H, P(E) is the spectral family of H, one has $f(H)dP(E) = f(E)dP(E)$.

An element (m,n) of f(H) is the (m,n) entry of f(J)

$$\begin{aligned}(f(J))_{m,n} &= < U_m | f(H) | U_n > = \int_\Sigma f(E)d < U_m | P(E) | U_n > \\ &= \int_\Sigma f(E)d < U_0 | p_m(H)\, P(E) p_n(H) | U_0 > \\ &= \int_\Sigma f(E)p_m(E)p_n(E) \quad d < U_0 | P(E) | U_0 >.\end{aligned}$$

Therefore, the spectral decomposition of a Jacobi matrix is

(1.2) $$(f(J))_{m,n} = \int_\sigma f(E)p_m(E)p_n(E)dN_0(E),$$

where N_0 is a nondecreasing function, σ is the support of N_0, the set of points E such that $N_0(E + \varepsilon) > N_0(E - \varepsilon)$ for any $\varepsilon > 0$, it could be only a part of Σ. The simplest function f is $f(x) = 1$, in which case f(J) is the identity matrix, showing immediately

$$\int_\alpha p_m(E)p_n(E)dN_0(E) = \delta_{m,n},$$

that the polynomials p_n are the orthonormal polynomials with respect to dN_0. This shows that the matrix J is completely determined by N_0 (*inverse spectral* problem), through the construction of the related orthogonal polynomials and their recurrence relation ([8], [15],etc...).

Another important function f is $f(x) = (z - x)^{-1}$, giving a Green operator element

(1.3) $G_{m,n}(z) = (z - J)^{-1}_{m,n} = \int_\sigma (z - E)^{-1} p_m(E)p_n(E)dN_0(E)$, $z \notin \sigma$.

Let us define

(1.4) $$q_n(z) = G_{0,n}(z).$$

One has

(1.5) $$q_n(z) = p_n(z)G_0(z) - p^{(1)}_{n-1}(z),$$

where $G_0(z) = G_{0,0}(z)$ and $p^{(1)}_{n-1}(z) = \int_\sigma \frac{p_n(z) - p_n(E)}{z - F} dN_0(E)$ is a polynomial of degree n - 1. The functions $q_n(z)$ and $p^{(1)}_{n-1}(z)$ (written $Q_n(z)$ in [17]) are also solutions of the recurrence relations (1.1), but with $(z - a_0)q_0(z) - b_1q_1(z) = 1$. Consequences are

(1.6) $$p_nq_{n-1} - p_{n-1}q_n = p_{n-1}p^{(1)}_{n-1} - p_np^{(1)}_{n-2} = 1/b_n.$$

The interest of $q_n(z)$ is that it is the *minimal* solution of (1.1) [15] : one has $q_n(z)/p_n(z) \xrightarrow[n\to\infty]{} 0$ when $z \notin \sigma$. Behaviours for large z are

$$p_n(z) \sim z^n/(b_1 \ldots b_n) \quad , \quad q_n(z) \sim b_1 \ldots b_n z^{-n-1}, \ |z| \gg 1.$$

One has also

(1.7) $$G_{m,n}(z) = p_m(z)q_n(z) \text{ if } m \leqslant n, \ p_n(z)q_m(z) \text{ if } m \geqslant n$$

indeed, if $m \leqslant n$, one has just to add to (1.3) the vanishing integral $\int_\sigma \frac{p_m(z) - p_m(E)}{z - E} p_n(E)dN_0(E)$ (p_n is orthogonal to polynomials of degree $< n$). See also chapter 12 of [39] for more on $(z - J)^{-1}$.

The behaviour of (1.3) near σ solves theoretically the *direct spectral* problem (find N_0 from J)

$\text{Im}\, G_{m,n}(E + i\varepsilon) \xrightarrow[\varepsilon>0]{\varepsilon\to 0} -\pi\, p_m(E)p_n(E)n_0(E)$, almost everywhere on σ where $n_0(E)$, the local density of states for $|U_0>$ is the derivative of the absolutely continuous part of N_0 ([16], [17]).

Relations for N_0 itself can also be found (Chebyshev-Markov inequalities [2] chap. 2,§5, [8] p.70, [14] §3).

The *average* density of states of an operator ([4],[17]§23, [40]) can also be studied through a Jacobi matrix [14] constructed from moments, or modified moments [21]. Non selfadjoint operators present further problems [16], [41].

2. Examples of asymptotic behaviour.

Numerous studies on special orthogonal polynomials and their weight functions ([8] chap. 6) give precious indications on the variation of the coefficients.

a) Chebyshev polynomials of the first kind.

$$n_0(E) = (1 - E^2)^{-1/2}/\pi \qquad -1 \leqslant E \leqslant 1 \qquad a_n = 0 \quad b_1 = 1/\sqrt{2} \quad b_n = 1/2,\ n > 1$$

$$G_0(z) = q_0(z) = (z^2-1)^{-1/2},\ q_n(z) = \sqrt{2}(z^2-1)^{-1/2}\left[z-(z^2-1)^{1/2}\right]^n,$$

$$p_n(E) = \sqrt{2}\, T_n(E),\ n \geqslant 1.$$

The values 0 and 1/2 reached by a_n and b_n are related to the support [-1,1] . This observation will be commented in the subsequent examples. If the support is $[E_1, E_2]$, the corresponding values are $(E_1 + E_2)/2$ and $(E_2 - E_1)/4$.

b) Chebyshev polynomials of the second kind. Bernstein-Szegö polynomials.

$$n_0(E) = 2(1 - E^2)^{1/2}/\pi \qquad -1 \leqslant E \leqslant 1 \qquad a_n = 0 \qquad b_n = 1/2 \qquad n \geqslant 1$$

$$p_n(E) = U_n(E) \qquad q_n(z) = 2\,[\, z-(z^2-1)^{1/2}\,]^n, \qquad n \geqslant 0.$$

This slight change in b_1 shows that boundedness properties of n_0 are not indicated by asymptotic behaviour. It will be shown in §5 that $a_n = 0$, $b_n = 1/2$ for $n > \nu_0$ corresponds to $n_0(E) = (1 - E^2)^{1/2}/R(E)$, where R is some polynomial of degree $\leqslant 2\nu_0$ (Bernstein-Szegö case).

c) Jacobi polynomials.

$$n_0(E) = c^t(1 - E)^\alpha(1 + E)^\beta \qquad -1 \leqslant E \leqslant 1 \qquad p_n(E) = c^t\, P_n^{(\alpha,\beta)}(E)$$

$$a_n = (\beta^2 - \alpha^2)/\,[\,(2n+\alpha+\beta)(2n+\alpha+\beta+2)\,] \sim (\beta^2 - \alpha^2)/(4n^2)$$

$$b_n = \{4n(n+\alpha)(n+\beta)(n+\alpha+\beta)/\,[\,(2n+\alpha+\beta-1)(2n+\alpha+\beta)^2(2n+\alpha+\beta+1)\,]\}^{1/2}$$

$$\sim \frac{1}{2} - \frac{2\alpha^2 + 2\beta^2 - 1}{16n^2}.$$

The amplitude of the perturbation of order n^{-2} in a_n and b_n gives the exponents of $n_0(E)$ at the band-edges, up to their signs (see appendix of [21]). This phenomenon will be encountered in §9.

Conversely, consequences of a law $a_n \sim c/n^2$, $b_n \sim 1/2 + d/n^2$ have been studied [9] : if $d + c/2$ and $d - c/2$ are $< 1/16$, N_0 has only a finite number of jumps outside $[-1,1]$; if one of these numbers is large enough, an infinite number of jumps occur outside $[-1,1]$.

d) The weight function $n_0(E) = c^t |E|^\alpha (1 - E^2)^\beta$.

The corresponding orthogonal polynomials are related to the Jacobi polynomials ([8] p.156) by

$$p_{2n}(E) = c^t P_n^{(\beta,(\alpha-1)/2)}(2E^2 - 1), \quad p_{2n+1}(E) = c^t E P_n^{(\beta,(\alpha+1)/2)}(2E^2 - 1)$$

with $a_n = 0$, $b_{2n} = \{4n(n+\beta)/[(4n+2\beta+\alpha-1)(4n+2\beta+\alpha+1)]\}^{1/2}$

$$b_{2n+1} = \{(2n+\alpha+1)(2n+2\beta+\alpha+1)/[(4n+2\beta+\alpha+1)(4n+2\beta+\alpha+3)]\}^{1/2}$$

i.e., $b_n \sim \frac{1}{2} - (-1)^n \frac{\alpha}{4n} + \frac{1 - 4\beta^2 + (-1)^n 2\alpha(2\beta+\alpha)}{16n^2}$.

This is our first example of interior singularity. The rate of damping is n^{-1}, with an oscillation related to the position of the singularity (see §7 and §9). In this case, the stability of asymptotic behaviour has been established : if $n_0(E)/|E|^\alpha$ is an even continuously derivable positive function on $[-1,1]$, $a_n = 0$ and $b_n = \frac{1}{2} - (-1)^n \frac{\alpha}{4n} + o(n^{-1})$ ([30] p.127). One observes that the non oscillating part of the perturbation is $(1 - 4\beta^2)/(16n^2)$, related to what happens at the band edges, independent of what happens elsewhere : a very favourable situation, that we hope to be valid in more complicated cases.

e) Pollaczek polynomials.

Up to now, only n^{-2} non oscillating perturbations have been encountered. The Pollaczek polynomials ([8] chap.6,§5) family allows n^{-1} behaviour :

$$p_n(E) = c^t P_n^{(\lambda)}(E;a,b,c),$$

$$a_n = -b/(n + \lambda + a + c) \sim -\frac{b}{n} + \frac{b(\lambda+a+c)}{n^2}$$

$$b_n = \{(n+c)(n+2\lambda+c-1)/[4(n+\lambda+a+c)(n+\lambda+a+c-1)]\}^{1/2}$$

$$\sim \frac{1}{2} - \frac{a}{2n} + \frac{8a(a+\lambda+c-1/2)+1-4(\lambda-1/2)^2}{16n^2} .$$

The weight function is rather formidable

$$n_0(E) = c^t (\sin\theta)^{2\lambda-1} e^{(2\theta-\pi)t} |\Gamma(\lambda+c+it)|^2 |{}_2F_1(1-\lambda+it,c;c+\lambda+it;e^{2i\theta}|^{-2}$$

$$t = (aE+b)(1-E^2)^{-1/2} \quad , \quad \cos\theta = E$$

but the important feature is the behaviour near ± 1 ([30] §6.2)

$$n_0(E) \sim c^t \exp(-2\pi(a \pm b)(1-E^2)^{-1/2}) , \quad E \sim \pm 1$$
$$a > |b|$$

This is a case of *Lifshitz* behaviour ([23], see §9).

f) Several intervals.

When the support σ is made of several intervals (i.e., if there are gaps), the sequences $\{a_n\}$ and $\{b_n\}$ are periodic or quasi-periodic at the limit. See §6 for more details and [11], [37] for a complete description.

g) Examples of singular behaviour.

Since Anderson [3], operators with random disordered or quasi-periodic elements are believed to have singular spectra : pure point or singularly continuous. One-dimensional systems are well understood, by direct spectral methods ([4], with a discussion of Thouless and Aubry relations) and inverse methods [10] [5] .

h) Unbounded intervals.

Hermite-like (Freud) orthogonal polynomials related to

$n_0(E) = c^t |E|^\rho \exp(-|E|^\alpha)$ on the whole real line correspond to $a_n = 0$, $b_n \sim (\frac{n}{c(\alpha)})^{1/\alpha}$ with $c(\alpha) = \frac{2\Gamma(\alpha)}{(\Gamma(\alpha/2))^2}$ (very likely true for all $\alpha > 0$; see [27] for references to recent works).

This gives for Laguerre-like polynomials (same weight function on $E > 0$ only) $a_n \sim 2(\frac{n}{c(2\alpha)})^{1/\alpha}$, $b_n \sim (\frac{n}{c(2\alpha)})^{1/\alpha}$.

3. Equations relating $a_n - a_n^*$, $b_n - b_n^*$ and $n_0(E) - n_0^*(E)$.

The effect on the coefficients of a change in the weight function depends on the orthogonal polynomials p_n and p_n^* related to $n_0(E)$ and $n_0^*(E)$. A way to appreciate the difference is known as Bernstein's integral equation [32] found by expressing that $p_n = \sum_{k=0}^{n} \alpha_{k,n} p_k^*$, $\alpha_{k,n} = \int_\sigma p_n(E)p_k^*(E)n_0^*(E)dE$, is orthogonal to $p_0^*,\dots,p_{n-1}^*$ with respect to $n_0 : \alpha_{k,n} = \int_\sigma p_n(E)p_k^*(E)(n_0^*(E) - n_0(E))dE$, $k < n$, $\alpha_{n,n} = b_1^*\dots b_n^*/(b_1\dots b_n)$.

One finds

$$b_1^*\dots b_n^* p_n(z) = b_1\dots b_n p_n^*(z) + b_1^*\dots b_n^* \int_\sigma p_n(E)\Big(\sum_{k=0}^{n} p_k^*(E)p_k^*(z)\Big)(n_0^*(E) - n_0(E))dE. \tag{3.1}$$

Comparison of coefficients of degrees n and n - 1 in

$p_n(z) = (b_1\dots b_n)^{-1} [z^n - (a_0+\dots+a_{n-1})z^{n-1} +\dots]$ and $p_n^*(z)$ yields

$$\frac{b_1^*\dots b_n^*}{b_1\dots b_n} - \frac{b_1\dots b_n}{b_1^*\dots b_n^*} = \int_\sigma p_n(E)p_n^*(E)(n_0^*(E)-n_0(E))dE$$

$$a_0^* + \dots + a_{n-1}^* - (a_0 +\dots+a_{n-1}) = \frac{b_1\dots b_n}{b_1^*\dots b_{n-1}^*} \int_\sigma p_n(E)p_{n-1}^*(E)(n_0^*(E)- n_0(E))dE.$$

If $n_0(E)$ is an infinitesimal perturbation of $n_0^*(E)$, we have as first approximation

$$\begin{aligned} b_n - b_n^* &\sim \frac{b_n^*}{2} \int_\sigma ((p_n^*(E))^2- (p_{n-1}^*(E))^2)(n_0(E)- n_0^*(E))dE \\ a_n - a_n^* &\sim b_n^* \int_\sigma (p_{n+1}^*(E)p_n^*(E) - p_n^*(E)p_{n-1}^*(E))(n_0(E) - n_0^*(E))dE. \end{aligned} \tag{3.2}$$

The point of view of *scattering techniques* can also be useful [7].

4. Continued fractions techniques; terminators.

As $q_n(z)$ is a solution of the recurrence relation (1.1), with $(z-a_0)q_0 - b_1q_1 = 1$, one has

$$q_0(z) = (z-a_0-b_1q_1(z)/q_0(z))^{-1}, q_n(z)/q_{n-1}(z) = b_n(z-a_n-b_{n+1}q_{n+1}(z)/q_n(z))^{-1}$$

or

$$G_0(z) = (z-a_0-b_1G_1(z))^{-1} , G_n(z) = b_n(z-a_n-b_{n+1} G_{n+1}(z))^{-1} \tag{4.1}$$

if $G_n(z)$ is defined as $q_n(z)/q_{n-1}(z)$. $G_0(z)$, and its imaginary part $-\pi\, n_0(E)$ if $z = E + i\varepsilon$, can then be computed by repeated application of this rational transformation (continued fraction) with indexes in decreasing order, from some reasonable estimation of $G_n(z)$ (terminator, or tail [38]).

The rational transformation relating directly G_0 and G_n can be obtained by writing back $q_n = p_n\, G_0 - p^{(1)}_{n-1}$:

(4.2)
$$G_n = \frac{p_n\, G_0 - p^{(1)}_{n-1}}{p_{n-1} G_0 - p^{(1)}_{n-2}}$$

whose inversion, using (1.6), gives

$$G_0 = \frac{p^{(1)}_{n-1}}{p_n} + \frac{1}{b_n p_n (p_n/G_n - p_{n-1})} ,$$

showing that the usual continued fraction approximant $p^{(1)}_{n-1}/p_n$ corresponds to the crudest estimate $G_n = 0$ for the terminator. Much better results will be obtained with terminators coming from model continued fractions with a nonvanishing imaginary part on σ.

Before exploring the simplest of these models, let us remark that $G_n(z)$ is not the same thing as $G_{n,n}(z)$, a diagonal element of the Green operator, whereas $G_n(z)$ (written $G^{>}_{n,n}$ in [37]) corresponds to an operator *starting* with $b_n, a_n, b_{n+1}, \ldots$. The difference is demonstrated by the expansions in powers of z^{-1}, interesting on their own right :

$$G_n(z) = b_n\, z^{-1} + a_n b_n z^{-2} + b_n(a_n^2 + b_{n+1}^2) z^{-3} + \ldots$$

(4.3)
$$G_{n,n}(z) = z^{-1} + a_n z^{-2} + (b_n^2 + a_n^2 + b_{n+1}^2) z^{-3} + \ldots$$

The coefficient of z^{-k-1} in $G_{n,n}(z)$ is a diagonal element of J^k (as $G(z) = (z - J)^{-1} = \sum_0^\infty J^k z^{-k-1}$); for $G_n(z)/b_n$, one has just to ignore $b_n, a_{n-1}, b_{n-1}, a_{n-2}, \ldots$.

5. The square-root terminator.

Let us consider the continued fraction expansion of a root of the quadratic equation $RG_0^2 - SG_0 + V = 0$, where R, S and V are real polynomials.

If the root is written

$$G_0(z) = (S(z) - X^{1/2}(z))/(2R(z)), \tag{5.1}$$

$$X(z) = S^2(z) - 4R(z)V(z) = (z-E_1)(z-E_2)\dots(z-E_{2m}),$$

$E_1 < E_2 \dots < E_{2m}$ also assumed to be real, G_0 will have a nonvanishing imaginary part on the region where $X(z) < 0$, which is the set made of m intervals

$$\sigma = [E_1, E_2] \cup [E_3, E_4] \cup \dots \cup [E_{2m-1}, E_{2m}].$$

In order to discuss this imaginary part, one looks carefully at $X^{1/2}(z)$, continuous outside σ, behaving like z^m when $z \to \infty$. One finds that the argument of $X^{1/2}(E + i\varepsilon)$, $\varepsilon > 0$, is 0 if $E > E_{2m}$, $\pi/2$ for $E_{2m-1} < E < E_{2m}$, π for $E_{2m-2} < E < E_{2m-1}$ (the rightmost gap), etc... . As

$$n_0(E) = (\operatorname{Im} X^{1/2}(E + i\varepsilon))/(2\pi R(E)), \quad E \in \sigma, \tag{5.2}$$

the weight function will be positive on σ if R has an odd number of sign changes in each gap.

Poles of G_0 are avoided if S interpolates $X^{1/2}$ at the zeros of R. Conversely, a fairly large family of weight functions with support σ can be approximated by an expression of the form (5.1).

After this capacity of approximation, the next interesting feature of the form (5.1) is its reproducibility under the transformation (4.1) :

if $S_n^2 - 4R_n V_n = X$,

$$G_n = \frac{S_n - X^{1/2}}{2R_n} = \frac{2V_n}{S_n + X^{1/2}} = b_n \left[z - a_n - b_{n+1} \frac{2(z-a_n)V_n/b_n - S_n - X^{1/2}}{2b_{n+1} V_n/b_n} \right]^{-1} \tag{5.3}$$

so that

$$R_{n+1}(z)/b_{n+1} = V_n(z)/b_n$$

$$S_{n+1}(z) = 2(z-a_n)V_n(z)/b_n - S_n(z) \tag{5.4}$$

$$b_{n+1} V_{n+1}(z) = b_{n+1} \frac{S_{n+1}^2(z) - X(z)}{4R_{n+1}(z)} = (z-a_n)(S_{n+1}(z) - S_n(z))/2 + b_n R_n(z).$$

Moreover, the degrees of R_n, S_n and V_n will decrease and stabilise ultimately on the values m-1, m and m-1. Indeed, by putting (5.3) in (4.2), one obtains R_n and S_n in terms of $p_n, p_{n-1}, p_{n-2}^{(1)}$, and, after transformations using (1.6)

(5.5) $R_n/b_n = X^{1/2}p_{n-1}q_{n-1} + Rq_{n-1}^2$, $S_n = X^{1/2}b_n(p_{n-1}q_n+p_nq_{n-1})+2Rb_nq_{n-1}q_n$.

As $q_n(z)$ behaves like $b_1\ldots b_n z^{-n-1}$ when $z \to \infty$, the degrees of R_n and S_n will decrease by steps of 2 units, downto m-1 and m.

Another way to write (5.5) is, from (1.6) and (1.7),

(5.6) $R_n/b_n = X^{1/2}G_{n-1,n-1} + Rq_{n-1}^2$, $S_n = X^{1/2}(1+2b_nG_{n-1,n})+2Rb_nq_{n-1}q_n$.

This allows

- *to check* if the square root model (5.1) is well adapted to a given set of a_n's and b_n's : the product of the expansions of

$X^{1/2}(z) = z^m - \frac{1}{2}(E_1 +\ldots+ E_{2m})z^{m-1} + \frac{1}{8}(2\sum_{k<l} E_kE_l - \Sigma E_k^2)z^{m-2}\ldots$ and

$G_{n-1,n-1}$ (see (4.3)) or $1 + 2b_nG_{n-1,n}$ should be close to polynomials, i.e., the coefficients of $z^{-1}, z^{-2},\ldots$ of the products should tend towards zero when $n \to \infty$.

- *to build* a first approximation of the terminator G_n by keeping the coefficients of the non negative powers of z :

$$R_n(z)/b_n \sim z^{m-1} + (a_{n-1} - \frac{1}{2}(E_1 +\ldots+ E_{2m}))z^{m-2}\ldots$$

(5.7) $$S_n(z) \sim z^m - \frac{1}{2}(E_1+\ldots+E_{2m})z^{m-1} + (2b_n^2+\frac{1}{8}(2\sum_{k<l} E_kE_l - \Sigma E_k^2))z^{m-2}\ldots$$

- *to recover* a_{n-1} and b_n from the preceding expansions, if R_n and S_n are generated by (5.4).

The easiest case occurs when m = 1 (single band), where one has $R_n(z)/b_n \equiv 1$ and $a_{n-1} = (E_1 + E_2)/2$,

$S_n(z) = z - (E_1 + E_2)/2$ and $b_n = (E_2 - E_1)/4$, if n is large enough. The corresponding orthogonal polynomials are known as the Bernstein-Szegö polynomials [8] p.204. Partically, $(E_1+E_2)/2$ and $(E_2-E_1)/4$ must be the limits of a_n and b_n when $n \to \infty$, a very well known situation.

There are other ways of obtaining square-root estimates of the form (5.1) for G_0, *especially if the band edges are not accurately known.* Quadratic Padé approximation seems a very promising approach [28] ; see also [36] for a technique assuming exponentially damped perturbations on a_n and b_n (not asymptotically correct but manageable).

6. Asymptotic behaviour of orthogonal polynomials.

If $E_1 = -1$, $E_2 = 1$, the square root terminator reduces to $G_n(z) = z - (z^2-1)^{1/2}$. On $z = E + i\varepsilon$, $-1 \leq E \leq 1$, this can be written $G_n(E + i\varepsilon) = e^{-i\theta}$, with $E = \cos\theta$. The product $q_n(z) = G_0(z)\ldots G_n(z)$ is then expected to behave like $e^{-in\theta}$ on σ, and the orthogonal polynomial $p_n(E)$ as $\sin n\theta$, from the imaginary part of $q_n(E + i\varepsilon)$. This is the basis of the description of the oscillatory character of $p_n(E)$, which will be used in the formulas of §3.

Before going into a more accurate description, including phase and envelope description, let us show that this behaviour still holds in the multiband case, knowing that the sequences of coefficients of R_n and S_n, together with the sequences $\{a_n\}$ and $\{b_n\}$, appear to be periodic or quasi-periodic [11], [37].

Let us consider now a consequence of (5.3) and (5.4) :

$$(6.1) \qquad (G_n \ldots G_{n+r-1})^2 = \frac{S_n - X^{1/2}}{S_n + X^{1/2}} \cdots \frac{S_{n+r-1} - X^{1/2}}{S_{n+r-1} + X^{1/2}} \; \frac{R_{n+r}/b_{n+r}}{R_n/b_n} .$$

If r is a period or a quasi period, $R_{n+1}/b_{n+r} \sim R_n/b_n$, and when $z = E+i\varepsilon$, $E \in \sigma$, $|G_n \ldots G_{n+r-1}| \sim 1$: let $G_n(E + i\varepsilon)\ldots G_{n+r-1}(E + i\varepsilon) \sim e^{-ir\theta}$.

So, we have again a variable θ, *which depends only on* σ. We refer to [11], [37] for the complete derivation (see also [20],[29], [32]), stopping at the form $G_n(E + i\varepsilon)\ldots G_{n+r-1}(E + i\varepsilon) \sim T_r(E) - U_{r-m}(E)X^{1/2}(E + i\varepsilon)$, where $T_r(E) = \cos r\theta$ is a polynomial of degree r in E ([37] §3.3, where it is called $P_p(\lambda)$). The uniform envelope of $T_r(E)$ on σ (see figure 9 of [37]) implies a differential equation $r\, Y^2(E)(T_r^2(E) - 1) = X(E)(T_r'(E))^2$, where the polynomial Y of degree m-1 takes into account the zeros of T_r' in each gap. For θ, this means

$$(6.2) \qquad \theta = i \int_E^{E_{2m}} Y(E)\, X^{-1/2}(E + i\varepsilon)dE, \qquad \theta(E_{2m}) = 0, \quad \theta(E_1) = \pi.$$

The supplementary conditions $T_r(E_{2k}) = T_r(E_{2k+1})$, or $\theta(E_{2k}) = \theta(E_{2k+1})$, $k=1,\ldots,m-1$, determine completely the monic polynomial Y. For instance, to $X(E) = E(E-5)(E-6)(E-10)$ corresponds $Y(E) = E - 5.5026$ ([26] §5).

A consequence of (6.2) is that $\theta'(E)$ is *infinite* at the band-edges :

(6.3) $$\theta(E) \sim \theta(E_k) - c^t\, |E - E_k|^{1/2} \quad \text{near } E_k,\ E \in \sigma.$$

The conditions of approximate periodicity are now explained : one must have $T_r(E_{2k}) \sim T_r(E_{2k+1}) \sim \pm 1$, i.e., the numbers $r\,\theta(E_{2k})$, $k=1,\dots,m-1$, must all be close to integer multiples of π. By taking indexes separated by such exact or approximate periods, one observes therefore that, as in the case of a single band, $q_n(E+i\varepsilon) = G_0(E+i\varepsilon)\dots G_n(E+i\varepsilon)$ shows an $e^{-in\theta}$ factor in its main behaviour. A consequence is that the zeros of $p_n(E)$ correspond to multiples of π/n for θ : the limit density of zeros of the orthogonal polynomials $iY(E)Y^{-1/2}(E+i\varepsilon)/\pi$, $E \in \sigma$. As the density is also the average, or global density of states of the operator J [14],

$$iY(E)X^{-1/2}(E+i\varepsilon)/\pi = -\pi^{-1} \lim_{r\to\infty} r^{-1} \sum_{n=0}^{r-1} \operatorname{Im} G_{n,n}(E+i\varepsilon)$$

or

(6.4) $$Y(E) = \lim_{r\to\infty} r^{-1} \sum_{n=0}^{r-1} R_n(E)/b_n, \quad \text{from (5.6).}$$

Up to now, we have described periodicity or quasi-periodicity behaviour occuring in the asymptotic regime, with its consequences on the oscillatory character of p_n and q_n. For a more complete description ([1], [32]), let us consider (1.5), $q_n = G_0\dots G_n$, and (6.1) for $(G_0\dots G_n)^2$, in order to show that $q_n(z)$ has no pole (if there is no Dirac peak in n_0), and may vanish only at some of the zeros of R_{n+1}, where $S_n(\rho_{k,n+1}) = -S_{n+1}(\rho_{k,n+1}) = X^{1/2}(\rho_{k,n+1})$ (see (5.3) and (5.4)). We consider now the integral of $(z-t)^{-1}X^{-\frac{1}{2}}(t)\log q_n(t)\,dt$ on a closed curve starting at $E_1 + \varepsilon i$, joining $+\infty + \varepsilon i$, making a big circle, and returning to $E_1 - \varepsilon i$ through $+\infty - \varepsilon i$. With the exception of $t = z$, the curve does not enclose singular points, we obtain $2\pi i$ times the residue at $t = z$, or $-2\pi i X^{-1/2}(z)\log q_n(z)$. On the other hand, along σ, summing the contributions above and below the real axis (because $X^{-1/2}(E-i\varepsilon) = -X^{-1/2}(E+i\varepsilon)$ on σ), one has to integrate $(z-E)^{-1}X^{-1/2}(E+i\varepsilon)\log|q_n(E+i\varepsilon)|^2 = (z-E)^{-1}X^{-1/2}(E+i\varepsilon)\log(R_{n+1}(E)/b_{n+1}R(E))$ (from (6.1)). In the gaps, $\log q_n(E+i\varepsilon)$ and $\log q_n(E-i\varepsilon)$ differ by an integer multiple of $2\pi i$, which decreases by $2\pi i$ when E crosses a zero of

q_n, whose value is $2\pi i(n+1)$ when $E > E_{2m}$, as $\log q_n(z) \sim -(n+1)\log z$ for large z. These two evaluations of the same integral are equated :

$$(6.5)\quad \log q_n(z) = X^{1/2}(z)\,\{-(2\pi i)^{-1}\int_\sigma (z-E)^{-1}X^{-1/2}(E+i\varepsilon)\log(R_{n+1}(E)/b_{n+1}R(E))\,dE - \sum_{k=1}^{n-1}\nu_{k,n}\int_{E_{2k}}^{E_{2k+1}} + \sum_{\text{zeros }\rho_{k,n+1}}\int^{E_{2k+1}} - (n+1)\int_{E_{2m}}^{\infty}(z-E)^{-1}X^{-1/2}(E)\,dE\} - (n+1)\pi i.$$

Remembering the connection between R and n_0 in our model (5.2), the result (6.5) establishes a direct relation between the weight function n_0 and asymptotic indications, writing the integral on σ as

$$(6.6)\qquad \int_\sigma (z-E)^{-1}X^{-1/2}(E+i\varepsilon)\,\log\left(\frac{|R_{n+1}(E)|}{b_{n+1}}\,2\pi\,|X^{-1/2}(E)|\,n_0(E)\right)dE.$$

This is therefore exact if $X^{1/2}(E)/n_0(E)$ is a polynomial (extended Bernstein-Szegö case). However, if $\int_\sigma |X^{-1/2}(E)|\log n_0(E)\,dE > -\infty$ (Szegö's condition), the integral (6.6) has a meaning and can be expanded in powers of z^{-1}. The m-1 first coefficients of this expansion allow the determination of the integers $\nu_{1,n},\ldots,\nu_{m-1,n}$ and R_{n+1} (Jacobi inversion problem [1], [32]). This defines the asymptotic terminator, including a_n and b_n, as hyperelliptic functions of arguments in arithmetic progression with n [1], [11], [20], [32] , [37].

Taking (6.5) with (6.6) as a valid asymptotic formula (a complete proof has been given by Szegö ([35] chap. 12, see also [6]) in the single band case; see [1], [32] for progress in the general case), one finds for the real part :

$$\mathrm{Re}\,\log q_n(E+i\varepsilon) \sim \frac{1}{2}\log(2\pi\,|X^{-1/2}(E)|\,n_0(E)\,|R_{n+1}(E)|/b_{n+1})$$

or

$$|q_n(E+i\varepsilon)| \sim [2\pi n_0(E)\,|R_{n+1}(E)|\,b_{n+1}]^{1/2}|X(E)|^{-1/4},\quad E\in\sigma.$$

The main term in the imaginary part of (6.5) has already been recognized as $-n\theta$, from our previous first explorations. The final asymptotic form can therefore be written

$$q_n(E+i\varepsilon) \sim [2\pi n_0(E)\,|R_{n+1}(E)|/b_{n+1}]^{1/2}\,|X(E)|^{-1/4}\,e^{-in\theta(E)-i\varphi(E)-i\tau_n(E)}$$

where $\tau_n(E)$ is periodic or quasi periodic in n, and

$$\varphi(E) = -(2\pi)^{-1}X^{1/2}(E+i\varepsilon) \;⨍_{\sigma} (E-E')^{-1}\, X^{-1/2}(E'+i\varepsilon)\log(2\pi n_0(E')|R_{n+1}(E')|\, |X^{-1/2}(E')|\, b_{n+1})dE'.$$

The orthonormal polynomial $p_n(E)$ is obtained by

$\mathrm{Im}\, q_n(E+i\varepsilon) = p_n(E)\, \mathrm{Im}\, G_0(E+i\varepsilon) = -\pi\, n_0(E)\, p_n(E)$:

(6.7) $p_n(E) \sim (2\,|R_{n+1}(E)|/b_{n+1})^{1/2}(\pi\, n_0(E))^{-1/2}\, |X(E)|^{-1/4}\sin(n\theta+\varphi+\tau_n)$.

Summary of §5 and §6.

If $n_0(E)$ is $\geqslant 0$ on the set of intervals σ where $X(E) < 0$, and regular enough in order to have $\int_\sigma |X(E)|^{-1/2}\log n_0(E)dE > -\infty$ (Szegö's condition), the form (5.3) is an asymptotically valid terminator. If σ is made of m intervals, the degrees of R_n and S_n are m-1 and m; S_n interpolates $\pm X^{1/2}(E)$ at the zeros of R_n (one in each gap). These polynomials can be estimated from diagonal Green operator elements (5.6). The resulting oscillatory behaviour of the orthononormal polynomials p_n is mainly given by an angular variable $\theta(E)$ depending only on σ (6.2), which can be computed by accumulating the products (6.1) on an approximated period.

7. Van Hove singularities.

By this name, one means singularities $0 < n_0(E) < \infty$ (and, perhaps, logarithmic singularities). For instance, near a point E_0, $n_0(E)$ behaves like $n_0(E_0) + c^t|E - E_0|^\alpha$, $\alpha > 0$, or n_0 is discontinuous at E_0, with $0 < n_0(E_0 - 0)$, $n_0(E_0 + 0) < \infty$, etc... The estimate (6.7) can then be put in (3.2) and yield interesting results. Taking θ as a natural variable, $a_n - a_n^*$ and $b_n - b_n^*$ appear as *Fourier series coefficients*. By this method, Hodges [18] has succeeded in settling the proper Van Hove singularities problem.

In general, let us consider (3.2) with the estimate (6.7) for $p_n^* \sim p_n$. The integrals will contain rapidly oscillating functions $\cos((2n+k)\theta + 2\varphi + \tau_{n+k} + \tau_{n+1})$, $-1 \leqslant k,l \leqslant 1$, and $n_0(E) - n_0^*(E) \sim \mu(E-E_0)$ creating the singularity at a point E_0. The integral of such a product can therefore be estimated by a Fourier coefficient, taking θ as variable of integration :

(7.1) $$\mu(E-E_0)dE \sim \nu(\theta-\theta_0)d\theta \quad , \quad \theta_0 = \theta(E_0).$$

The change of variable is very different according as E_0 is interior to a band, or is a band edge : in the first case $\mu(E-E_0)dE \sim \frac{dE_0}{d\theta_0}\mu(\frac{dE_0}{d\theta_0}(\theta-\theta_0))d\theta$, in the second case, from (6.3),

$E \sim E_0 + c^t(\theta-\theta_0)^2$, $\mu(E-E_0)dE \sim 2\, c^t(\theta-\theta_0)\mu(c^t(\theta-\theta_0)^2)d\theta$.

The effect of a singularity $\nu(\theta-\theta_0)$ on a Fourier coefficient for large n is, in most cases, a damped oscillation with a change of phase

$$\int_0^\pi f(\theta)\nu(\theta-\theta_0)\cos(n\theta+\xi)d\theta \sim f(\theta_0)\cos(n\theta_0+\xi+\eta)\rho(n).$$

Here are two examples [22] § 3.5 and 5.5

$\nu(\theta-\theta_0)$	η	$\rho(n)$
0 if $\theta < \theta_0$, $A(\theta-\theta_0)^\alpha$ if $\theta > \theta_0$	$(\alpha+1)\pi/2$	$A\Gamma(\alpha+1)/n^{\alpha+1}$
$A(\theta_0-\theta)^\alpha$ if $\theta < \theta_0$, 0 if $\theta > \theta_0$	$-(\alpha+1)\pi/2$	$A\Gamma(\alpha+1)/n^{\alpha+1}$.

Application to integrals of the form $\int_\alpha p_{n+k}(E)p_{n+1}(E)(n_0(E)-n_0^*(E))dE$ gives $\pi^{-1}[|R_{n+k+1}(E_0)R_{n+l+1}(E_0)|/(b_{n+k+1}b_{n+l+1})]^{1/2}(n_0(E_0))^{-1}|X(E_0)|^{-1/2}\rho(2n)$

$$\cos[(2n+k+l)\theta_0 + 2\varphi_0 + \tau_{n+k}(E_0) + \tau_{n+l}(E_0) + \eta),$$

where η and $\rho(n)$ are related to $\nu(\theta-\theta_0)$. This can be translated back in products of combinations of p_{n+k} and q_{n+k}, p_{n+l} and q_{n+l}.

One can therefore describe and simulate the effect of one or several Van Hove singularities by the following rules :

1) The main behaviour of a_n and b_n is given by a_n^* and b_n^*, generated by an asymptotically valid square-root terminator (5.3).

2) To a_n^* and b_n^*, one must add damped oscillations as in Fourier coefficients of functions with singularity $\nu(\theta-\theta_0)$, related to the singularity $\mu(E-E_0)$ of $n_0(E)$ by (7.1). If E_0 is interior to a band, ν and μ are similar, if E_0 is a band edge, $\nu(\theta-\theta_0) = 2c^t(\theta-\theta_0)\mu(c^t(\theta-\theta_0)^2)$. For instance, if the

singularity is $|E - E_0|^{\alpha}$, $\alpha \geq 0$, $\nu(\theta-\theta_0)$ is as $|\theta - \theta_0|^{\alpha}$ or $|\theta - \theta_0|^{2\alpha+1}$ respectively, giving damping rates of $n^{-\alpha-1}$ or $n^{-2\alpha-2}$. See [24] for a striking connection.

3) The oscillations of the perturbations are given by $\frac{1}{2} b_n^*(y_n^2 - y_{n-1}^2)$ for $b_n - b_n^*$, $b_n^*(y_{n+1}y_n - y_n y_{n-1})$ for $a_n - a_n^*$, where y_n is some solution of the recurrence relation (1.1) in the asymptotic range for n, i.e., some combination of $p_n^*(E_0)$ and $q_n^*(E_0)$ (or $p_{n-1}^{(1)*}(E_0)$).

Things are simpler when there is only one band : the oscillations are described by $\sin((2n-1)\theta_0 + 2\varphi_0 + \eta)$ and $\sin(2n\theta_0 + 2\varphi_0 + \eta)$ [18], [25], [17] §21.

4) The effect of several singularities is the sum of the effects of each singularity. This holds for the main asymptotic contribution; nonlinear mixing of modes appear if one wants to go further, for instance by iterating the correction in (3.2).

For non Van Hove singularities (i.e., when $n_0(E_0) = 0$ or ∞), the rule 2) is no more valid, mainly because the estimate (6.7) fails near E_0 : the orthogonal polynomial cannot be infinite nor vanish always at E_0. One must therefore look for more refined estimates of orthogonal polynomials, but the study of a new class of model functions [12], $[32]_2$, [26] will give directly information on the recurrence coefficients. Moreover, an interesting new family of terminators will also appear.

8. The Riccati terminator.

The widely used square-root terminator (5.3) is exact for weight functions with square root singularities at the band edges, and asymptotically exact for most regular weight functions (up to a finite number of singularities). A better understanding, and, perhaps, the possibility of better numerical algorithms, is given by terminators satisfying a Riccati equation

(8.1) $$AG_n' = B_n G_n^2 + C_n G_n + D_n$$

where A_n, B_n, C_n and D_n are polynomials. By direct inspection, or by writing G_n as a ratio of solutions of second order linear differential equations ([19] §12.52), the non polar singularities of G_n appear to be the zeros of A. An interesting connection with the square root terminator is got by writing (8.1) as

$$G_n = [-C_n - (C_n^2 - 4B_n(D_n - AG_n'))^{1/2}] / (2B_n)$$

$$\sim [-C_n - (C_n^2 - 4B_n D_n)^{1/2}] / (2B_n) + A(C_n^2 - 4B_n D_n)^{-1/2} G_n'$$

near a zero of A. The differential equation for G_n is then approximately linear, solved by

$$G_n(z) \sim c^t \exp \int^z [C_n^2(t) - 4B_n(t)D_n(t)]^{1/2} A^{-1}(t)dt + \text{particular sol.}$$

The same behaviour (*exact* if $B_n = 0$) holds for the imaginary part of $G_n(E + i\varepsilon)$, i.e., for $n_0(E)$ with $n = 0$.

If E_0 is a simple root of A, $n_0(E)$ will behave like $| E - E_0 |^{\alpha}$ (algebraic, or Jacobi singularity) with $\alpha = [C_0^2 - 4B_0D_0]^{1/2}/A'$ at E_0; if the multiplicity of E_0 is $p+1 > 1$, $n_0(E) \sim c^t \exp(- \alpha(E-E_0)^{-p})$ with $\alpha = (p+1)(p-1) ! [C_0^2 - 4B_0D_0]^{1/2}/A^{(p+1)}$ at E_0 (Lifshitz singularity).

The next important property of the Riccati model is that its form is kept under the elementary rational transformation (4.1). One finds [26]

(8.2)
$$B_{n+1}/b_{n+1} = D_n/b_n$$
$$C_{n+1} = -C_n - 2(z-a_n)D_n/b_n$$
$$b_{n+1} D_{n+1} = A + b_nB_n + (z-a_n)C_n + (z-a_n)^2D_n/b_n.$$

An important consequence is

(8.3) $$C_n^2 - 4B_nD_n = C_0^2 - 4B_0D_0 - 4A(D_0/b_0 + \ldots + D_{n-1}/b_{n-1}).$$

The Riccati terminator should match asymptotically the square root terminator, the connection has been made clear by Gammel & Nuttall [12], comparing (5.4)(8.2)

(8.4)
$$D_n \sim - 2n\ Z\ V_n$$
$$C_n \sim 2n\ Z\ S_n$$
$$B_n \sim - 2n\ Z\ R_n$$
$$C_n^2 - 4B_n D_n \sim 4n^2 Z^2 X$$

where Z is some rational function. Moreover,

$$C_n^2 - 4B_nD_0 = -4A(D_0/b_0+\ldots+D_{n-1}/b_{n-1}) \sim 4n^2AZ \text{ (average of } V_k/b_k)$$
$$\sim 4n^2AZY \quad \text{from (6.4), so that}$$

(8.5) $$ZX = AY$$

Z is the polynomial vanishing at the zeros of Y (in the gaps), and at the zeros of A which are *not* zeros of X, i.e., at the *interior singularities*. One can then study more accurately the influence of the singularities on the terminator, and therefore on a_n and b_n.

9. Algebraic and Lifshitz singularities.

A behaviour $n_0(E) \sim c^t |E - E_0|^\alpha$, $\alpha > -1$, near $E = E_0$, can be produced by the Riccati model (8.1) if E_0 is a simple root of A and with $C_0^2 - 4B_0D_0 = \alpha^2A'^2$ at E_0. As the values of $C_n^2 - 4B_nD_n$ are independent of n on the zeros of A (from (8.3)), this polynomial takes fixed *positive* values at the zeros of A, whereas its general behaviour is given by $4n^2Z^2X$, strongly negative on σ. This results in a set of small neighbourhoods of the singular points where $C_n^2 - 4B_nD_n$ is positive, whose length are about n^{-1} at the interior singular points (double roots of $Z^2X = 0$), n^{-2} at the band edges (single roots of $Z^2X = 0$) : *the Riccati terminator (8.1) behaves like a square root terminator with supplementary small gaps of lengths of about n^{-1}, and band edges displaced towards the interior of the bands by amounts of about n^{-2}*. As the effect on a_n and b_n of small gaps [37] and band edges displacements are known, we conclude to n^{-1} and n^{-2} perturbations, with oscillations very likely similar to what happens with Van Hove singularities. A study of the corresponding orthogonal polynomials (generalized Jacobi polynomials [12], [31]) shows indeed that the estimate (6.7) fails in the indicated neighbourhoods ([31] p. 675,h)). For more quantitative work, one must solve accurately (8.2) and (8.3), starting with (8.4). For instance [12], with $C_n(z) \sim 2nZ(z)S_n(z) + \gamma_n(z)$, one obtains recurrence relations for the values of γ_n at the zeros of A...

Lifshitz behaviour is treated on similar lines : a Lifshitz singularity E_0 is a zero of multiplicity p+1 of A. From (8.5), the multiplicity in Z is p+1 or p, according as E_0 is interior or band-edge; 2p + 2 or

$2p+1$ for Z^2X. Near E_0, $C_n^2 - 4B_nD_n$ behaves therefore like $c - dn^2(E-E_0)^{2p+2}$ or $c-dn^2(E-E_0)^{2p+1}$, giving apparent small gaps of lengths $\sim n^{-1/(p+1)}$ and apparent band edges displacements of $n^{-1/(p+1/2)}$.

As an example of more accurate treatment, let us consider an even weight function behaving like $\exp(-\alpha/(1-E^2)^p)$ near ± 1. Then, $C_n^2 - 4\,B_nD_n \sim 4\alpha^2p^2$ near $\pm$ 1 (where $A \sim (1-E^2)^{p+1}$). Assuming very slow variation of a_n, b_n, (8.2) gives $C_n \sim C_{n+1} \sim z\,D_n/b_n$, and (8.3) : $C_n^2 - 4B_nD_n \sim (z^2-4b_n^2)(D_n/b_n)^2 \sim C_0^2 - 4B_0D_0 + O((1-E^2)^{p+1})$, so that D_n/b_n is approximately the Taylor expansion of $2\alpha p(E^2-4b_n^2)^{-1/2}$ about $E^2 = 1$, limited to degree $2p$

$$D_n/b_n \sim 2\alpha p \sum_{k=0}^{p} \frac{\Gamma(k+1/2)}{k!\,\Gamma(1/2)} (1-4b_n^2)^{-k-1/2} (1-E^2)^k .$$

Equating the coefficients of $(1-E^2)^{2p+1}$ in $(E^2-4b_n^2)(D_n/b_n)^2 \sim 4n^2Z^2X$, with $Z = (E^2-1)^p$ and $X = E^2 - 1$, gives $1 - 4b_n^2 \sim \left(\frac{(p-1)!\,\Gamma(1/2)}{\alpha\Gamma(p+1/2)}\,n\right)^{-1(p+1/2)}$, or

$$b_n \sim \frac{1}{2} - \frac{1}{4}\left(n\,\frac{(p-1)!\,\Gamma(1/2)}{\alpha\Gamma(p+1/2)}\right)^{-1(p+1/2)} .$$

Comparison with values computed by Ph. Lambin for $\exp(-2(1-E^2)^{-1})$ ($\alpha=2, p=1$) is of the highest interest :

n	exact b_n	estimated $b_n \sim \frac{1}{2} - \frac{1}{4}n^{-2/3}$
10	0.450929	0.44613
15	0.461700	0.45889
20	0.467983	0.46606.

The estimate has been made for integer values of p, but comparison with Pollaczek's result (§2) suggests that it could be true for any real $p > 0$ (replacing $(p-1)!$ by $\Gamma(p)$).

Similar work with $n_0(E) = e^{-\alpha E^{-p}}$, $-1 \leqslant E \leqslant 1$, p even, gives estimates of the form $b_n \sim \frac{1}{2} - (-1)^n\,c^t\,n^{-1/(p+1)}$ ...

Conclusions

Some aspects of the variations of a_n and b_n with respect to the weight function $n_0(E)$ are summarized in the table nearby.

Feature of n_0	Behaviour of a_n and b_n	Mathematical status
Single band $[E_1, E_2]$	$a_n \underset{n\to\infty}{\longrightarrow} (E_1+E_2)/2$ $b_n \underset{n\to\infty}{\longrightarrow} (E_2-E_1)/4$	Completely proved by Szegö ([35] chap. 12) if $\int_{E_1}^{E_2} [(E-E_1)(E_2-E)]^{-1/2} \log n_0(E)\,dE > -\infty$ Known to be true in some other cases [30].
Several bands support σ $[E_1, E_2]$ ... $[E_{2m-1}, E_{2m}]$	Limit behaviour inagreement with a square-root terminator [11] $\frac{S_n - X^{1/2}}{2 R_n}$ $X(z) = \prod_{k=1}^{2m} (z-E_k)$	Proposed condition $\int_\sigma |X(E)|^{-1/2} \log n_0(E)\,dE > -\infty$ (extension of Szegö's cond.) Rigorous results in [1], [32].
Singularity at $E_0 \in \sigma$ Nature of singularity	Supplementary damped oscillations to previous a_n^* and b_n^* : $a_n \sim a_n^* + b_n^* \rho_n (y_{n+1} - y_{n-1}) y_n$ $b_n \sim b_n^* + \frac{1}{2} b_n^* \rho_n (y_n^2 - y_{n-1}^2)$ y_n is some solution of $b_{n-1}^* y_{n+1} = (E_0 - a_n^*) y_n - b_n^* y_{n-1}$. ρ_n is the rate of damping. Examples : ρ_n if E_0 interior; ρ_n if E_0 band edge	Hoped to be valid in the multiband case.

Van Hove $\|E-E_0\|^{\alpha}$	$n^{-1-\alpha}$	$n^{-2-2\alpha}$	[18] [25] (interior, one band)
Algebraic $\|E-E_0\|^{\alpha}$	n^{-1}	n^{-2}	progress in [12][31][32]
Lifshitz $\exp(-c^t\|E-E_0\|^{-p})$	$n^{-1/(p+1)}$	$n^{-1/(p+1/2)}$	p=1/2 : Pollaczek (see §2) (band-edge, one band)

Acknowledgements.

It is a pleasure to thank D.G. Pettifor, conference organizer; J.P. Gaspard and Ph. Lambin for their kind interest and precious information.

1 N.I. AHIEZER Orthogonal pol. on several intervals. Soviet Math. *1*(1960)989-992. With Yu. TOMČUK : ibid. *2*(1961)687-690.

2 N.I. AKHIEZER The Classical Moment Problem. Oliver & Boyd, Edinburgh 1946.

3 P.W. ANDERSON The Fermi glass-theory and experiment. Comments Solid State Phys. *2*(1970)193-198.

4 J. AVRON, B. SIMON Almost periodic Schrödinger operators II. The integrated density of states. Duke Math. J. *50*(1983)369-391.

5 M. BARNSLEY, D. BESSIS, P. MOUSSA The Diophantine moment problem... J. Math. Phys. *20*(1979)535-546. D. BESSIS, M.L. MEHTA, P. MOUSSA Orthogonal pol. on a family of Cantor sets... Letters Math. Phys. *6*(1982)123-140.

6 W. BARRETT An asymptotic formula...J. London Math. Soc. (2) *6*(1973) 701-704.

7 K.M. CASE Scattering theory, orthogonal pol... J. Math. Phys. *15*(1974)974-983; 2166-2174.

8 T.S. CHIHARA An Introduction to Orthogonal Pol. Gordon & Breach, N.Y., 1978.

9 T.S. CHIHARA Orthogonal pol. whose distribution functions have finite point spectra. SIAM J. Math. An. *11*(1980)358-364.

10 W. CRAIG Pure point spectrum...Commun. Math. Phys. *88*(1983)113-131.

11 F. DUCASTELLE, P. TURCHI, G. TRÉGLIA Band gaps and asymptotic behaviour ...These Proceedings.

12 J.L. GAMMEL, J. NUTTALL Note on generalized Jacobi pol., pp. 258-270 in D. & G. CHUDNOVSKY, editors, Lecture Notes Math. 925, Springer 1982.

13 J.P. GASPARD Generalized moments methods...These Proceedings, See also in C. BREZINSKI et al, eds, Lecture Notes Math., Springer 1985.

14 J.P. GASPARD, F. CYROT LACKMANN Density of states from moments.... J. Phys. C *6*(1973)3077-3096.

15 W. GAUTSCHI Minimal solutions...Math. Comp. *36*(1981)547-554; On generating orthogonal pol. SIAM J. Sci. Stat. Comp. *3*(1982)289-317.

16 G. GROSSO, G. PASTORI PARRAVICINI chap. III and IV to appear in special volume of Adv. Chem. Phys. "Memory function approaches...".

17 R. HAYDOCK The recursive solution of the Schrödinger eq., pp. 215-294 in H. EHRENREICH et al., eds. : Solid State Phys. 35(1980); these Proceedings.

18 C.H. HODGES Van Hove singularities...J. Physique Lett. *38*(1977) L187-L189.

19 E.L. INCE Ordinary Differential Equations, Longmans Green 1927 = Dover.

20 Y. KATO On the spectral density of periodic Jacobi matrices, pp. 153-181 in M. JIMBO, T. MIWA, eds. Proceedings RIMS Symposium ... World Science, Singapore, 1983.

21 Ph. LAMBIN, J.P. GASPARD Continued fraction technique...Phys. Rev. B *26*(1982)4356-4368.

22 M.J. LIGHTHILL Introduction to Fourier Analysis and Generalized Functions. Cambridge U.P. 1958.

23 I.M. LIFSHITZ The energy spectrum of disordered systems. Adv. In Phys. *13*(1964)483-536.

24 D.S. LUBINSKY, P. RABINOWITZ Rates of convergence of Gaussian quadrature for singular integrands. To appear in Math. Comp.

25 Al. MAGNUS Recurrence coefficients for orthogonal pol...pp.150-171 in L. WUYTACK, editor, Lect. Notes Math. 765, Springer, 1979.

26 Al. MAGNUS Riccati acceleration..., pp. 213-230 in H. WERNER, H.J. BÜNGER eds. Lect. Notes Math. 1071, Springer, Berlin, 1984.

27 Al. MAGNUS A proof of Freud's conjecture..., to appear in C. BREZINSKI et al., eds., Lect. Notes Math.

28 I.L. MAYER, J. NUTTALL, B.Y. TONG Quadratic Padé approximant method for calculating densities of states. Phys. Rev. B*29*(1984)7102-7104.

29 P. van MOERBEKE, The spectrum of Jacobi matrices. Inv. Math. 37(1976) 45-81.

30 P. NEVAI Orthogonal Polynomials, Memoirs AMS n°213(1979).

31 P. NEVAI Mean convergence of Lagrange interpolation III. Trans. AMS *282*(1984)669-698.

32 J. NUTTALL, S.R. SINGH Orthogonal pol... on a system of arcs, J. Approx. Th. *21*(1977)1-42. J. NUTTALL Asymptotics of diagonal

Hermite-Padé pol. to appear in J. Approx. Th.

33 O. PERRON Die Lehre von den Kettenbrücken. Teubner 1929 = Chelsea.

34 M. REED, B. SIMON : Methods of Modern Mathematical Physics I. Ac. Press 1972.

35 G. SZEGÖ Orthogonal Polynomials, AMS, Providence 1939.

36 A. TRIAS, M. KIWI, M. WEISSMANN Reconstruction of the density of states from its moments. Phys. Rev. B *28*(1983)1859-1863.

G. ALLAN A linear predition of recursion coefficients. J. Phys. C *17*(1984)3945-55; these Proceedings.

37 P. TURCHI, F. DUCASTELLE, G. TREGLIA Band gaps and asymptotic behaviour of continued fraction coefficients. J. Phys. C *15*(1982)2891-2924.

38 H. WAADELAND Tales about tails. Proc. AMS *90*(1984)57-64.

39 H.S. WALL Analytic Theory of Continued Fractions. Van Nostrand 1948.

40 F. WEGNER Bounds on the density of states in disordered systems. Z. Phys. B *44*(1981)9-15.

41 H. STAHL Orthogonal pol. of complex valued measures...pp. 771-788 in G. ALEXITS, P. TURÁN eds : Fourier Analysis and Approximation Theory vol. II. North-Holland, Amsterdam, 1978.

POSTSCRIPT.

According to work done by Rakhmanov (Math. USSR Sbornik 32 (1977) 199-213), commented and simplified by A. Mate, P. Nevai and V. Totik (Constructive Approximation 1 (1984) 63-69, extended by P. Nevai (to appear in C. Brezinski et al., editors: Orthogonal Polynomials and Applications, Springer Lecture Notes in Math. Series), the condition

$$n_0(E) > 0 \qquad \text{almost everywhere in } [E_1, E_2]$$

is <u>sufficient</u> to imply

$$a_n \to a_\infty = (E_1 + E_2)/2, \qquad b_n \to b_\infty = (E_2 - E_1)/4 \qquad \text{when } n \to \infty.$$

The result is not altered if δ functions (isolated if outside $[E_1, E_2]$) are added to n_0.

Band Gaps and Asymptotic Behaviour of Continued Fraction Coefficients

F. Ducastelle, P. Turchi[1], and G. Tréglia[2]

Office National d'Etudes et de Recherches Aérospatiales (ONERA), B.P. 72
F-92322 Chatillon Cédex, France

1. Introduction

A very efficient and now widely known method of approximating densities of states is to use continued fraction expansions (for a review see [1]). As far as the applications to solid state physics problems are concerned, the method has been first applied to densities of states with connected supports (i.e. without gaps), in which case the coefficients of the continued fraction converge. In the presence of gaps, they exhibit undamped oscillations, whose characteristics are now well understood. The first detailed study of these oscillations is due to MAGNUS [2]. Many of his results have also been obtained by the present authors, using techniques more familiar to the physicist [3]; in particular, the problem of determining the asymptotic behaviour of the coefficients has been shown to reduce to that of studying the spectrum of periodic linear chains. This problem in turn has been discussed in some detail in another context by TODA [4]. It is obviously not necessary to repeat all the arguments contained in these references and we shall just recall the main stages of our own work. More recent developments can be found in MAGNUS' contribution.

We start from a continued fraction f(z) written as

$$f(z) = \frac{1\,|}{|z-a_1} - \frac{b_1^2\,|}{|z-a_2} - \ldots\ldots - \frac{b_n^2\,|}{|z-a_{n+1}} - \ldots\ldots \tag{1}$$

With this continued fraction is associated a Jacobi tridiagonal matrix which may also be viewed as the matrix representation of a (e.g. tight-binding) hamiltonian H on a semi-infinite linear chain (1,....,n,...), with

$$H_{n,n} = a_n \; ; \; H_{n,n+1} = b_n \tag{2}$$

It is convenient to define the resolvent $G^{1>}(z) = (z-H)^{-1}$ whose matrix element $G^{1>}_{n}$ is then identical with f(z). We shall also use chains beginning at site n ; the corresponding resolvent (or Green function) is noted $G^{n>}$, and its matrix element $G^{n>}_{nn}$ is given by

[1]Also at : Laboratoire de Dynamique du Réseau et Ultra-Sons
Université Pierre et Marie Curie, Tour 22,
4 Place Jussieu, 75005 Paris, FRANCE

[2] Present address : Laboratoire de Physique des Solides, Bât. 510
Université Paris-Sud, Centre d'Orsay,
91405 Orsay, FRANCE

$$f_n(z) = G_{nn}^{n>}(z) = \cfrac{1}{z-a_n - \cfrac{b_n^2}{z-a_{n+1} - \cdots}} \tag{3}$$

In the case of connected spectra, $a_n \to a_\infty$, $b_n \to b_\infty$, so that $f_n \to f_\infty$, with

$$f_\infty(z) = [\, z - a_\infty - b_\infty^2 f_\infty(z)]^{-1} \tag{4}$$

and since $f_\infty(z)$ should behave as $1/z$ when $z \to \infty$, the correct solution is

$$f_\infty(z) = 2/\; (z-a_\infty) + [(z-a_\infty)^2 - 4b_\infty^2]^{1/2} \tag{5}$$

which corresponds to a spectrum centred on a_∞ and of width $4b_\infty$.

In the presence of gaps, it can be argued, and it is "experimentally" observed, that the coefficients become periodic functions of n when $n \to \infty$ (see Fig. 1 and [3]). We consider here the case of a single gap; we have then four band edges $E_1 < E_2 < E_3 < E_4$, and it is often convenient to use also the following quantities

$$2a = E_1 + E_4 \;;\; 2g = E_2 + E_3 \;;\; 2W = E_4 - E_1 \;;\; 2G = E_3 - E_2 \tag{6}$$

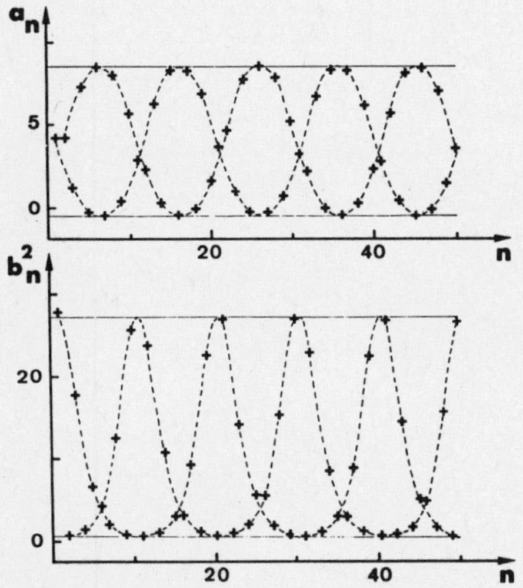

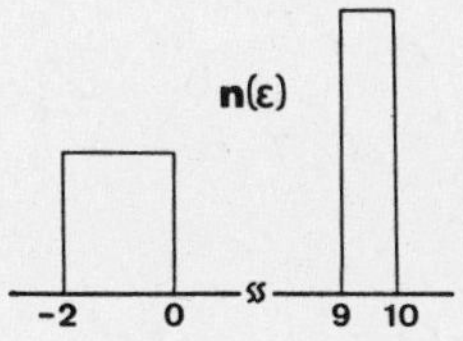

Fig. 1 Oscillating coefficients in the presence of a gap

In the case where both a_n and b_n have two limit points, a_1, a_2 and b_1, b_2 respectively, it is easy to calculate the two limits of $f_n(z)$ and the edges of the spectrum [5,3] ; in particular it is found that the gap is necessarily symmetric (g = a). In order to obtain an arbitrary gap, more general periodicities should be considered.

2. Periodic linear chains

Since we are interested in asymptotic properties, we may consider an infinite periodic linear chain with period p, $a_{n+p} = a_n$, $b_{n+p} = b_n$ and with periodic boundary conditions. Qualitatively, the spectrum of such a chain is well-known. Using Bloch's theorem, the eigenstates are characterized in part by a wave vector k varying between $-\pi/p$ and π/p. There are p branches E(k) which are the solutions of the secular equation

$$E^p - \sigma_1 E^{p+1} + \dots + (-1)^p \sigma_p - 2b_1 b_2 \dots b_p \cos pk = 0 \tag{7}$$

where all coefficients $\sigma_1, \dots, \sigma_p$ are independent of k [3]. From the hermiticity of H, we know that the p roots of (7) are real. Furthermore, these roots are simple in general, and since it is easily realized that the band edges correspond to $k = 0$ or $k = \pm \Pi/p$, (p-1) gaps are obtained in general. Thus, the condition that there is a single gap imposes severe conditions on the possible values of the independent coefficients $a_1, \dots, a_p$ and $b_1, \dots, b_p$. More precisely, we must impose that there are only four simple roots among all roots at $k = 0$ or $\pm \Pi/p$ (see Fig. 2). Making these conditions explicit provides us with recurrence laws for the coefficients and with explicit expressions for the matrix elements of Green function G(z) of the periodic chain [3]. In particular, $G_{nn}(z)$ is given by

$$G_{nn}(z) = (z - \alpha_n)/\sqrt{X(z)} \tag{8}$$

with

$$X(z) = (z-E_1)(z-E_2)(z-E_3)(z-E_4)$$

and

$$\alpha_n = \frac{1}{2}\sum_{i=1}^{4} E_i - a_n$$

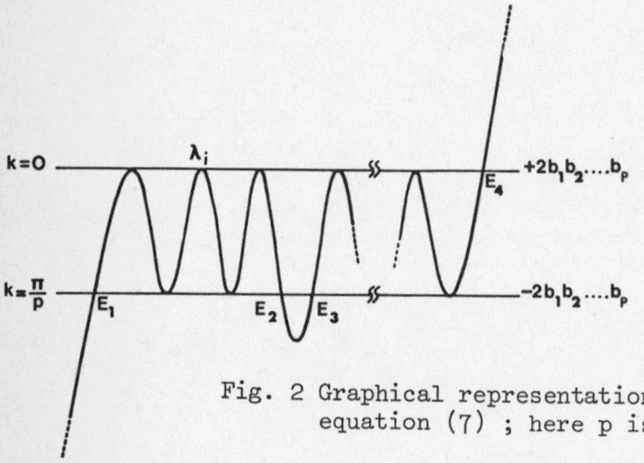

Fig. 2 Graphical representation of the secular equation (7) ; here p is odd

Different forms for the recurrence laws can be given. For example, if a_n and therefore α_n is given, b^2_{n-1} and b^2_n are obtained from

$$2b^2_{n-1} \ , \ 2b^2_n = -(\alpha_n^2 + A_1\alpha_n + A_2) \pm \sqrt{X(\alpha_n)} \tag{9}$$

with

$$A_1 = -\sum_i E_i/2 \ ; \ A_2 = \frac{1}{2}\sum_{i<j} E_i E_j - A_1^2/2$$

Conversely, if b_n is given, similar non-linear recurrence relations yield a_n and a_{n+1}. The final result is that if a coefficient is given, all others are determined, provided one also fixes an ordering assignment (e.g., one has to associate b^2_{n-1} and b^2_n with definite signs in the r.h.s. of (9)).

This may be visualized by using a phase space representation. Let (x,y) be the points with coordinates (b_{n-1}^2, a_n) or (b_n^2, a_n), then all points are on a single curve whose equation is easily obtained from (9). This curve is in fact the closed part of a cubic curve, an example of which is shown in Fig. 3. Then, it is easy to determine the extremal values of a_n and b_n^2 ; one finds that

$$a-G \leqslant a_n \leqslant a+G$$
$$\frac{W-G}{2} \leqslant b_n \leqslant \frac{W+G}{2} \tag{10}$$

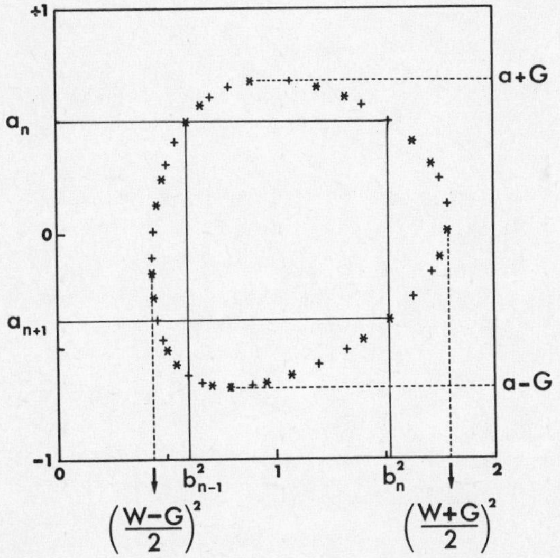

Fig. 3 Phase space representation of the recurrence laws; here a = 0, W = 2, g = 0.136 and G = 0.340

It should also be noted that the period determined by the recurrence laws is not a rational number in general, which means that, contrary to what has been assumed at the beginning of this section, the oscillations of the coefficients are generally not commensurate with the linear chain (this is clear in Fig. 1).

It turns out that in fact a_n and b_n^2 are periodic elliptic functions of n. Although this is not completely evident from the previous arguments, this could have been suspected from the appearance of $\sqrt{X(z)}$, i.e. of the square root of a polynomial of fourth degree [3]. A deeper argument is that the appropriate Riemann surface for describing a spectrum with a gap is precisely that involved in the study of elliptic functions. Then, it is easy to extend the analysis to the case of several gaps : the elliptic functions simply become abelian functions [2,6].

Note finally that in the case of a small gap, $G/W \ll 1$, a perturbation theory similar to that introduced before by HODGES [7] can be used ; then a_n and b_n become sinusoidal functions of n [3].

3. Termination of the continued fraction

A simple method for terminating a continued fraction is to insert at level n the asymptotic form of $f_n(z)$. Smoother procedures are certainly preferable and some of them are discussed in other contributions to this conference, but this method is not so bad, as shown below.

The first thing to do is to determine the asymptotic form of $G^{n>}$, i.e. to cut the periodic chain at position n. Using standard techniques [3], one finally finds two equivalent expressions for $f_n(z)$

$$f_n(z) = \frac{z^2+A_1 z + A_2 + 2b_{n-1}^2 - \sqrt{X(z)}}{2b_{n-1}^2\,(z+A_1+a_{n-1})} = \frac{2(z+A_1+a_n)}{z^2+A_1 z+A_2+2b_{n-1}^2 + \sqrt{X(z)}} \qquad (11)$$

These expressions involve two consecutive coefficients, but using the asymptotic recurrence relations, one coefficient and an ordering assignment suffice in principle. In practice, one should verify that the last known coefficients satisfy these relations reasonably well. Another problem is that to use (11) we need the band edges. They are not known in general, but as discussed elsewhere, several efficient methods for determining them can be devised [3].

To illustrate the importance of using an appropriate termination, we have calculated the coefficients of a very simple model density of states, made up of two rectangles. The constant termination $a_\infty = a$, $b_\infty = W/2$ is rather bad indeed whereas the termination (11) is much more satisfactory (Fig. 4). However, although this may look surprising from this figure, integrals over the density of states are almost independent of the termination if a large enough number of exact coefficients, about 20, is known.

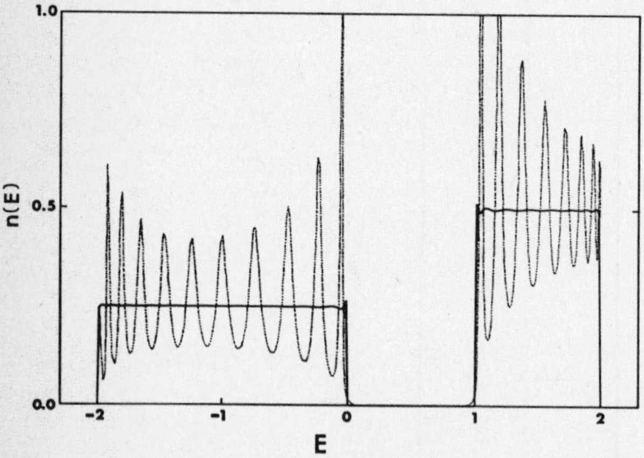

Fig. 4 Reconstruction of a rectangular density of states, using 20 exact levels. a = 0, W = 2, g = 0.5 and G = 0.5 - Dotted curve: constant termination; full curve: improved termination (11)

We note finally that all the arguments briefly sketched above can be extended to the case of several gaps, without any particular difficulty, in principle, if not in practice ! [2,3].

REFERENCES

1. R. Haydock: Solid State Phys. 35, 216 (1980)

2. A. Magnus: in Padé Approximation and its applications, Lecture notes in Mathematics 765, ed. L. Wuytack (Springer, Berlin 1979)

3. P. Turchi, F. Ducastelle and G. Tréglia: J. Phys. C. Solid State Phys. 15, 2891 (1982)

4. H. Toda: Theory of Nonlinear Lattices, Solid State Science Series nº 20 (Springer, Berlin 1981)

5. J.P. Gaspard and F. Cyrot-Lackmann: J. Phys. C. Solid State Phys. 6, 3077 (1973)

6. C.L. Siegel: Topics in Complex Function Theory, vol. II: Automorphic Functions and Abelian Integrals (Springer, Berlin 1980)

7. C.H. Hodges: J. Physique Lett. 38, L 187 (1977)

Computing Greenians: Quadrature and Termination

C.M.M. Nex[1]

Institute of Theoretical Science, University of Oregon, Eugene, OR 97403, USA

Permanent address: Cavendish Laboratory, Madingley Road
Cambridge CB3 OHE, United Kingdom

1 Introduction

Since the advent of the recursion method, its applications have been multitudinous [1], but the computation of the greenian remains an area of research aimed at improving the extraction of physical quantities from the results. The problem is to incorporate information about the analytic structure of the greenian, $G(e) = m(e) - i\pi n(e)$, with the explicitly calculated coefficients obtained in the application of the recursion algorithm to the given physical system. The principal line of enquiry [2] in this area has been oriented by the continued fraction representation of the greenian and analysis and extrapolation of its coefficients to take account of additional physical knowledge. The quadrature approach [3] has not proved fruitful in the incorporation of other than general analytic properties, but can in fact be so used as will be shown later.

In this paper I will first establish a notation and give a brief summary of general computational techniques using analytic termination and quadrature. In section 3 will be more details of a form of terminator which may incorporate quadrature calculations and more general types of specific analytic knowledge of the greenian. Finally I will present some ideas of matching a computed greenian with a model one, using a philosophy suggested by the quadrature formulation. These may be of use in estimating terminators from the computed recursion coefficients.

2 Recursion Method Formulation

A continued fraction representation of a greenian $G(e)$ is generated by the recursion method [1] from a hamiltonian H and starting state $\underline{u}_0$. Taking

$$b_0 = \underline{u}_0^T \underline{u}_0$$

and using the notation for real matrices we generate, for $i=0,1,2,..,n-2$

1 Supported by NSF-Condensed Matter Theory Grant DMR 81-22004.

$$a_i = \underline{u}_i^T H \underline{u}_i \, / \, \underline{u}_i^T \underline{u}_i \tag{2.1}$$

$$\underline{u}_{i+1} = (H - a_i) \underline{u}_i - b_i^2 \underline{u}_{i-1} \tag{2.2}$$

$$b_{i+1}^2 = \underline{u}_{i+1}^T \underline{u}_{i+1} \, / \, \underline{u}_i^T \underline{u}_i \tag{2.3}$$

In most of what follows there is an obvious dependence of expressions on n, the number of levels, which we hereafter leave implicit. Equations 2.1-3 are in a normalisation independent form , which enables any convenient normalisation to be used in computational practice. From this is derived [1] the continued fraction representation of the greenian

$$G(e) = \underline{u}_o^T (H - eI)^{-1} \underline{u}_0 = m(e) - i \pi n(e) \tag{2.4}$$

namely

$$G(e) = \cfrac{b_o^2}{e - a_0 - \cfrac{b_1^2}{e - a_1 - \cfrac{b_2^2}{\ddots \cfrac{b_{n-2}^2}{e - a_{n-2} - b_{n-1}^2 t(e)}}}} \tag{2.5}$$

In principle the continued fraction is infinite in length, but in practice is almost always truncated at some finite level and a termination function, t(e), introduced as in 2.5 . t(e) may be another continued fraction or some other analytic form or even zero.

The continued fraction is evaluated efficiently using the orthogonal polynomials $p_i(e)$ and $q_i(e)$ associated with the three-term recurrences related to 2.2:

$$p_{-1}=0 \; , \; p_0=1 \; , \; p_{i+1}(e) = (e-a_i) \, p_i(e) - b_i^2 \, p_{i-1}(e) \tag{2.6}$$

$$q_{-1}=0 \; , \; q_0=1 \; , \; q_{i+1}(e) = (e-a_{i+1}) \, q_i(e) - b_{i+1}^2 \, p_{i-1}(e) \tag{2.7}$$

Here the subscript corresponds to the degree of the polynomial (in some work it is one less than the degree for the polynomials of the second kind). Again we assume the polynomials will be arbitrarily but consistently normalised in computation and use expressions which are normalisation independent. The greenian 2.4 may then be evaluated [4] as

$$G(e) = \frac{q_{n-2}(e) - b_{n-1}^2 \; t(e) \; q_{n-3}(e)}{p_{n-1}(e) - b_{n-1}^2 \; t(e) \; p_{n-2}(e)} \quad . \tag{2.8}$$

This is reasonably fast to evaluate and can be conveniently programmed in real arithmetic if only the density of states is required [5].

In the quadrature approach [3] we focus on the imaginary part of the greenian and construct an algorithm for the evaluation of integrals over the local density of states n(e) defined in 2.4 . We write

$$F(e) = \int_{-\infty}^{e} f(x) \; n(x) \; dx \tag{2.9}$$

noting that the integrated density of states, N(e), is obtained by letting f(x)=1 . The expression approximating F(e) is

$$F(e) \cong \sum_{j=1}^{m-1} w(x_j(e)) \; f(x_j(e)) \; + z \; w(x_m(e)) \; f(x_m(e)) \quad . \tag{2.10}$$

z is a numerical value between 0 and 1 which may indeed be a function of e, as will become relevant later, but in the absence of any other information may be taken as 0.5 . The weights, $w(x_j(e))$, and nodes, $x_j(e)$, are generated as follows: a polynomial $p_n(x;e)$ is defined by first setting

$$a_{n-1} = e - b_{n-1}^2 \; p_{n-2}(e) \; / \; p_{n-1}(e) \tag{2.11}$$

and then using the recurrence 2.6 with x replacing e to define a polynomial, p_n, of order n in x but also depending on e. The nodes $x_j(e)$ are the zeros in x of $p_n(x;e)$, with one zero, $x_m = e$ by construction. The weights $w(x_j(e))$ are given by the expression

$$w(x_j(e)) = q_{n-1}(x_j(e)) \; / \; p'_n(x_j(e)) \quad . \tag{2.12}$$

This is a computationally convenient, normalisation independent form of the intuitively simpler identity which will be used later :

$$w(x) = \left[\sum_{j=0}^{n-1} (P^\circ_j(x))^2 \right]^{-1} \tag{2.13}$$

Here the $P^\circ_j(e)$ are the normalised polynomials proportional to $p_j(e)$ satisfying the symmetrised version of the recurrence 2.6.

$$p_j(e) = \prod_{i=0}^{j} b_i \; P^\circ_j(e) \tag{2.14}$$

The expression 2.10 for F(e) may be differentiated analytically with respect to e to give an approximation to such quantities as the local density of states as explicitly indicated in [4].

The quadrature formulation is considerably slower to evaluate than the terminated continued fraction, but requires no further input than the computed coefficients and is also less sensitive to errors in these. When f(x)=1 the extremal values of z in 2.10 provide rigorous lower and upper bounds on the integrated density of states [3].

3 Some Specific Forms of Terminator

As incorporation of analytic information into density of states calculations can improve the accuracy of computed results, there is considerable interest in terminators, t(e), which can be used to do this, and also in their estimation from the computed recursion coefficients. In this section we will outline some forms of these and also indicate a possible analagous procedure using the quadrature approach.

The form of terminator originally suggested [1] incorporated the knowledge of the band edges of a single band density of states, being the well-known 'square-root' terminator:

$$t(e) = [\, e - a - (e-a-2b)^{\frac{1}{2}} (e-a+2b)^{\frac{1}{2}} \,] \;/\; 2b^2 \qquad (3.1)$$

where a and b are the asymptotic values of a_i and b_i. This has been generalised [2],[5] to allow for several band gaps although still with 'square-root' band edges, to the form

$$t(e) = \frac{S_r(e) - X(e)}{T_{r-1}(e)} \quad \text{where} \qquad (3.2)$$

$$X(e) = \prod_{k=1}^{b} (e-\alpha_k)^{\frac{1}{2}} (e-\beta_k)^{\frac{1}{2}} \qquad (3.3)$$

for b bands in the intervals $[\alpha_k,\beta_k]$. $S_r(e)$ and $T_{r-1}(e)$ are polynomials of appropriate degree chosen either to match the analytic form of X(e) or the asymptotic coefficients a_i and b_i of 2.2 .

A recent suggestion [6] is for an alternative form which can be used to specify more precise analytic information such as the 'type' of the band edges or internal singularities when these are known. The terminator is derived from a model greenian, G*(e) , constructed to possess the desired analytic properties and whose continued fraction expansion may be readily computed. For example, for a density of states with several band gaps as in 3.2 with square-root band edges a model greenian could be

$$G^*(e) = \sum_{k=1}^{b} \frac{8N_k}{(\beta_k-\alpha_k)^2} \left[e - \frac{(\alpha_k+\beta_k)}{2} - (e-\alpha_k)^{\frac{1}{2}}(e-\beta_k)^{\frac{1}{2}}\right] \qquad (3.4)$$

where N_k is the weight of the kth band. The continued fraction expansion for such a $G^*(e)$ may be readily computed [4] as in the context of orthogonal polynomials, using the equations analagous to 2.1-3 . In table 1 are listed appropriate quadratures for some common types of features. The superposition of single band quadrature rules provides an appropriate composite inner product for the generation of the continued fraction corresponding to $G^*(e)$. If $p_i^*(e)$ and $q_i^*(e)$ are the polynomials corresponding to $p_i(e)$ and $q_i(e)$ in 2.6 and 2.7, but with respect to $G^*(e)$, the terminator is given by inversion of equation 2.8 (or 'unravelling' of the continued fraction for the greenian $G^*(e)$) :

$$t(e) = \frac{q^*_{n-2}(e) - G^*(e)\, p^*_{n-1}(e)}{b^{*2}_{n-1}\,[\, q^*_{n-3}(e) - G^*(e)\, p^*_{n-2}(e)\,]} \qquad (3.5)$$

Table 1 :
A list of greenians for some simple model bands. More general model greenians can be generated by the superposition of those listed, together with any others needed. The tridiagonalisation is generated using the union of the appropriate gaussian quadratures listed in the last column. N is the weight of the band; α and β are the band limits.

greenian	model	quadrature
$\frac{8N}{(\beta-\alpha)^2} \left[e - \frac{(\alpha+\beta)}{2} - (e-\alpha)^{\frac{1}{2}}(e-\beta)^{\frac{1}{2}}\right]$		Chebyshev (2nd. kind)
$\frac{N}{\pi} \left[\frac{1}{e-\alpha} - i\,\delta(e-\alpha) \right]$	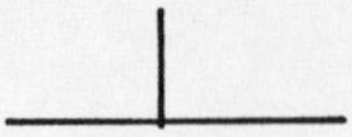	δ function
$\frac{N}{\pi(\beta-\alpha)} \left[\ln\left\lvert\frac{\alpha-e}{\beta-e}\right\rvert - i\,\Theta(e-\alpha)\,\Theta(\beta-e) \right]$		Legendre
$\frac{-4N}{(e-\alpha)^{\frac{1}{2}}(e-\beta)^{\frac{1}{2}}(\beta-\alpha)^2}$	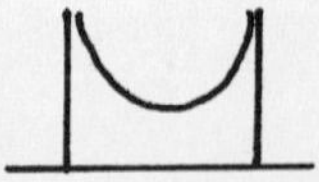	Chebyshev (1st.kind)

If the relative weight of each band is not known a priori, it may be adequately estimated by use of the quadrature approach. If $N^Q(e,z)$ is the quadrature estimate of the integrated density of states (2.10 with f(x)=1) at energy e and parameter z, then a good estimate of its value in the k th band gap is given by

$$N\left(\frac{\beta_k + \alpha_{k+1}}{2}\right) \cong \tfrac{1}{2}\ [\ N^Q(\beta_k,1) + N^Q(\alpha_{k+1},0)\ ] \quad . \quad (3.6)$$

This approach circumvents the problem of matching the phase of the oscillations in the continued fraction coeffients of G(e) and G*(e), as these phases seem to be determined by the relative weights of the bands [7].

A similar model greenian approach in the quadrature method could also provide a 'terminator' in the form of a function z(e) in 2.10 which imposes certain analytic conditions on $N^Q(e,z(e))$. A model N*(e) may be generated as outlined before with G*(e), and used to evaluate a z*(e) through inversion of 2.10 with f(x)=1 and all other terms known. This provides a non-decreasing approximation to N(e) between the rigorous upper and lower bounds, which, for instance, could have zero slope in the region of band gaps.

Of the three terminators outlined above, the new form based on model greenians offers the most scope for including analytic information in the recursion calculations. The time involved in the evaluation is not excessive in any of the continued fraction plus terminator methods, while the quadrature approach remains rather more expensive. If integrals are to be calculated, however, the latter does provide an effective solution as other numerical integration techniques are likely to be adversely affected by the singularities in the density of states. If necessary, an analytic terminator can be converted to a continued fraction form, and this appended to the original computed coefficients before applying the quadrature method.

4 Matching Greenians

We here discuss a use of the quadrature approach in estimating band limits of a computed continued fraction. Other methods include analysis of the continued fraction coefficients or of the behaviour of the extremal eigenvalues [8] with which the present suggestion needs to be compared. Rather than analysing the computed recursion coefficients, we here propose a matching of the behaviour of the weight w(e), 2.13, of the computed greenian G(e) with that of a model greenian, G*(e) and w*(e). We now make explicit the dependance of w(e) on n by the introduction of a subscript indicating the number of levels used in its computation, hence $w_n(e)$.

If, for instance, the number of distinct bands of a density of states is known, but not their precise location, the model greenian $G^*(e)$, 3.4, may be readily varied by changing α_k and β_k. The weight, $w_{n-1}(e)$, at the point of interest $e=x_m(e)$ can be calculated in a normalisation independent form derived using the Christoffel-Darboux identity [9] :

$$w_{n-1}(e) = \frac{p_{n-2}(e)\, q_{n-2}(e) - p_{n-1}(e)\, q_{n-3}(e)}{p_{n-2}(e)\, p'_{n-1}(e) - p'_{n-2}(e)\, p_{n-1}(e) + p^2_{n-1}(e)/b^2_{n-1}} \tag{4.1}$$

As noted in section 2, $w_n(e)$ is simplest expressed, 2.13, as the inverse norm of the truncated eigenvector (of length n and energy e) of the tridiagonalisation of the hamiltonian H. For greenians corresponding to the classical orthogonal polynomials we have the result [10] that, for e outside the band, $w_n(e)$ decays exponentially as a function of n, and as we have a determinant moment problem, we expect $w_n(e)$ to tend to zero within the bands, except perhaps at the singular points. Although the classical theorems have not been fully generalised to the many-band case, it seems from numerical experience that w_n decays to zero rapidly, perhaps exponentially, for energies

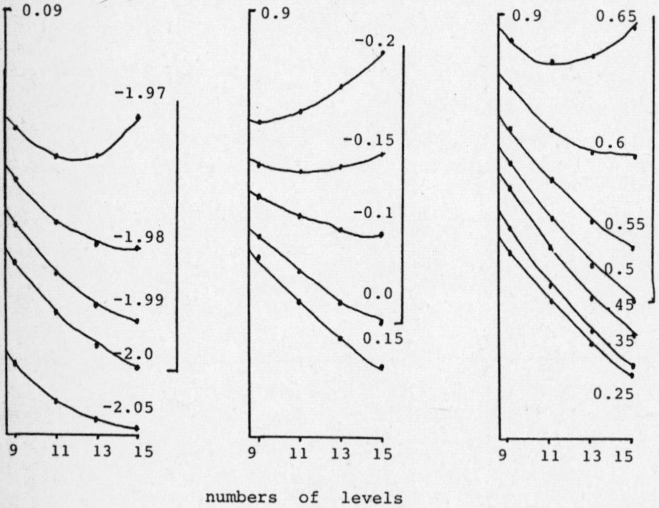

Fig.1

Plots of the decay of the scaled weight $n\, w_n(e)$ as a function of n for some specific energies in the vicinity of band edges, showing the increase in the decay rate outside the bands. The energies of evaluation, e, are indicated alongside each curve, those inside the bands are linked by a vertical brace. The density of states has 'square-root' band edges at -2.0, 0.0, 0.5 and 2.0.

in the band gaps and as 1/n within the bands. At delta functions in the density of states w_n tends to a constant while at band edges and Van Hove singularities its decay rate depends on the precise order of the singularity. For example, for the simple classical band edges $w_n \sim 1/n^3$ at a 'square root' type, $\sim 1/n^2$ at a 'logarithmic' type and $\sim 1/n$ at an 'inverse square root' type. The change in behaviour through a two-band density of states is illustrated in Fig.1 where $n\,w_n(e)$ is plotted as a function of n for values of e in the neighbourhood of the band edges.

The matching technique is an attempt to match the behaviour in the gap regions of the computed $w_n(e)$ with that of $w^*_n(e)$ corresponding to a model greenian, by varying the band edges of the model. This is in distinct preference to trying to identify the precise decay law in $w_n(e)$ and the variation thereof as a function of e. This circumvents the absence of a general theorem characterising the decay rate in general, and instead assumes, as is borne out by numerical experience, that the behaviour in the band gaps is dominated by the position of

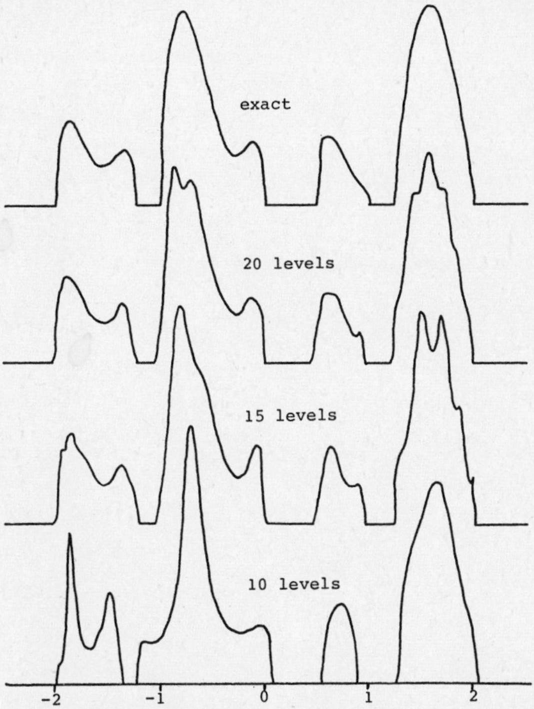

Fig.2

Approximations to a four-band density of states generated using the new terminator and estimating the band edges using the matching philosophy. 10, 15 and 20 level calculations are compared with the exact function.

the band edges and relative weights of the bands, not by the detailed structure within the bands. We show in Fig.2 the results of applying this technique to a known four-band density of states. The given information was the existence of four distinct bands and the recursion coefficients to the stated number of levels, and the program attempted to match the band-gap minima in $w_n(e)$ with those of a simple four-band set of semi-ellipses. The decrease in the extraneous structure as the number of levels increases is an obvious indication of the improvement in the estimation of the band edge locations with the increasing number of levels.

5 Conclusions and Acknowledgements

The recursion method continues to improve as a useful tool in the theoretical study of electronic and vibrational properties of materials. Greenian calculations are still developing from the early simplistic approaches, and much work remains to be done in evaluating and combining the many ideas in a fertile field.

I would like to acknowledge the many conversations with Roger Haydock, the hospitality of the Institute of Theoretical Science in the University of Oregon at Eugene, and the generosity of Cambridge University in granting me leave of absence, all of which contributed to the development of much of this work.

References

1 R.Haydock, V.Heine, D.Bullet and M.J.Kelly: Solid State Physics 35 (Academic, New York 1980)
2 P.Turchi, F.Ducastelle, and G.Treglia: J.Phys.C 15, 2891-2924 (1982)
3 C.M.M.Nex: J.Phys.A 11, 563-63 (1978)
4 R.Haydock and C.M.M.Nex: J.Phys.C to appear
5 C.M.M.Nex: Comp.Phys.Comm. to appear
6 A.Magnus: Pade Approximation and its applications, Lecture notes in mathematics 765 (Springer, Berlin 1979)
7 R.Haydock and C.M.M.Nex: to be published
8 T.S.Chihara: An Introduction to Orthogonal Polynomials (Gordon and Breach, New York 1978)
9 G.Szegö: Orthogonal Polynomials (Amer.Math.Soc., Providence 1967)

Application of Linear Prediction for Extrapolating Recursion Coefficients

G. Allan

Laboratoire de physique des solides*, Institut Supérieur d'Electronique du Nord, 3, Rue Francois Baës, F-59046 Lille Cédex, France

The linear prediction theory is used to extrapolate the recursion coefficients and to calculate band limits. The coefficients of the continued fraction are written as sums of complex exponentials. The exponential decays and amplitudes are obtained by a least square fit of the first known coefficients. Application is made to bulk Silicon s and p electronic densities of states. The agreement with an exact one obtained by integration over the Brillouin zone is very good.

1 INTRODUCTION

The recursion method is now widely used to calculate densities of states [1]. It can be applied to translationally invariant solids where it replaces the cumbersone calculation of energy levels for a large number of wavevectors and the integration over the Brillouin zone. But its main interest resides in its applicability to defect structures (vacancies, surfaces, dislocations for example) and to disordered materials. However, due to computing time one can calculate only a finite small number of recursion coefficients, so the density of states we get is only approximate. It does not present any Van Hove singularity and the density of states is small but neither exactly equal to zero inside the band gaps. Several attempts have been made to solve this problem and improve the continued fraction termination [2]. We here use the linear prediction theory to extrapolate the recursion coefficients. This procedure is based upon the fact that these coefficients show oscillations around values directly related to the band width and to the band center. Using the linear prediction theory, we assume that these oscillations can be approximated by a sum of complex exponentials. The exponential decays and amplitudes are then fitted to the first known recursion coefficients.

In section 3, this method is applied to the bulk Silicon s and p electronic states. We shall show that it leads to accurate band widths and to rough values of the gap limits. We also study the convergence of these limits with the number of exact coefficients.

2 LINEAR PREDICTIVE MODELLING OF THE RECURSION COEFFICIENTS

This method is more commonly used in digital signal processing [3]. Here we shall apply this method to extrapolate the recursion coefficients. Let us first recall the expression of the continued fraction

* Laboratoire associé au Centre National de la Recherche Scientifique.

$$G = \cfrac{1}{E - a_1 - \cfrac{b_1^2}{E - a_2 - \cfrac{b_2^2}{\cfrac{\cdot}{\cfrac{\cdot}{E - a_{l-1} - b_{l-1}^2 G_{ll}}}}}} \tag{1}$$

The density of states is simply equal to $-\mathrm{Im}\, G/\pi$. The a_n and b_n recursion coefficients are obtained by a tridiagonalization of the hamiltonian [4] or from the moments of the density of states [5]. Due to computation time, only a finite number of these coefficients can be calculated. Then the problem is to determine the continued fraction termination G_{ll}. For a long time, it has been shown that these coefficients oscillate around values a and b which are directly related to the band limits E_B and E_T [5]

$$b = \frac{(E_T - E_B)}{4} \tag{2.a}$$

$$a = \frac{(E_T + E_B)}{2} \tag{2.b}$$

Assuming that after the l-th level, the coefficients do not deviate too much from a and b, a first order perturbation theory is valid to calculate G_{ll} [6]. Let us call

$$\alpha_{2n-1} = \frac{a_n - a}{b} \tag{3.a}$$

$$\alpha_{2n} = \frac{2(b_n - b)}{b} \tag{3.b}$$

the reduced perturbation with respect to the constant values a and b. It has been shown that the α_n are equal to the Fourier components of the difference between the exact termination G_{ll} and the one we get assuming constant coefficients, g.

Within the band limits, let us recall that

$$g = \frac{E-a-i\sqrt{4b^2-(E-a)^2}}{2b^2} \tag{4}$$

Then we get in first order perturbation theory [6]

$$G_{ll} - g = \exp(-2i\phi) \sum_{n=0}^{\infty} \exp(-in\phi)\, \alpha_{n+2l-1} \tag{5}$$

where

$$\phi = \cos^{-1}[(E-a)/2b] \tag{6}$$

Close to a Van Hove singularity E_i, the density of states varies as $(E-E_i)^{\nu}$. Then for large n, the α_n vary like $n^{-(\nu+1)}\cos(n\phi_i+\Phi)$ [6]. Due to band gaps, the α_n present undamped oscillations [7-8]. As the band gaps are generally small compared to the band width, the first order perturbation theory is also valid. It shows that the oscillation amplitude of the α_n is proportional to the gap width. We here assume that in all the cases the α_n can be approximated by a sum of complex exponentials. This procedure cannot give the exact behaviour of the density of states close to Van Hove singularities or the exact gap limits when the first order perturbation theory is not valid. Nevertheless, we shall see that it can give a good estimation of the gap limits and very good values of the band limits. Generally the a and b values are unknown, but our procedure can also be applied in this case with only slight modifications.

First, let us consider this case where a and b are unknown. From equation (3), we can write

$$a_n = b\alpha_{2n-1} + a \tag{7.a}$$

$$b_n = \frac{b}{2}\alpha_{2n} + b \tag{7.b}$$

The coefficients $a_n/2$ and b_n can also be written as the odd and even terms of u_n

$$u_n = \frac{b\,\alpha_n}{2} + \frac{a+2b}{4} - \frac{a-2b}{4}(-1)^n \tag{8}$$

If we approximate α_n by a sum of exponentials, we also have

$$u_n = \sum_{m=1}^{M} C_m \exp(nx_m) \tag{9}$$

where x_m may be complex. The number M of exponentials will be discussed below.

We can equivalently write

$$u_n = \sum_{m=1}^{M} C_m r_m^n \qquad \text{with} \tag{10}$$

$$r_m = \exp(x_m) \tag{11}$$

If we compare (8) and (10), we see that one r_m is equal to unity. Then the corresponding C_m is equal to (a+2b)/4. Another r_m is equal to minus unity and the C_m to (2b-a)/4. We now consider the expression (10) valid for any n and we want to fit the C_m and the r_m to the known coefficients a_n and b_n.

Let us take the z-transform of the expression (10)

$$U(z^{-1}) = \sum_{n=0}^{\infty} u_n z^{-n} \tag{12}$$

$$= \sum_{m=1}^{M} \frac{C_m}{1-r_m z^{-1}} \qquad (13)$$

$U(z^{-1})$ can be expressed as the ratio of two polynomials $Q(z^{-1})$ and $P(z^{-1})$

$$U(z^{-1}) = \frac{Q(z^{-1})}{P(z^{-1})} \quad \text{where} \qquad (14)$$

$$Q(z^{-1}) = \sum_{m=0}^{M} q_m z^{-m} \qquad (15.a)$$

$$P(z^{-1}) = \sum_{m=0}^{M} p_m z^{-m} \qquad (15.b)$$

Let us here remark that the r_m are the roots of $P(z)$,

$$P(z^{-1}) * U(z^{-1}) = Q(z^{-1}) \qquad (16)$$

which is a convolution in the discrete domain.

This can be written

$$\sum_{m=0}^{M} p_m u_{n-m} = q_n \qquad 0 \leq n \leq M-1 \qquad (17.a)$$

$$\sum_{m=0}^{M} p_m u_{n-m} = 0 \qquad M \leq n \leq N \qquad (17.b)$$

where N is the number of known u_n which are related to the first recursion coefficients $a_n/2$ and b_n by the equations (7). The equation (17.b) is exact if the u_n are equal to a sum of exponentials. In our case this is only approximative and we look for a least-square solution. Equation (17.b) can be written as

$$\sum_{m=1}^{M} p_m u_{n-m} = -u_n \qquad (18)$$

as p_0 is equal to unity. Let us remark that this equation is the same as the one used by A. TRIAS et al. [9] to fit separately the a_n and b_n. We calculate the others coefficients p_m such as to minimize ε^2

$$\varepsilon^2 = \sum^{N} [u_n + \sum_{m=1}^{M} p_n u_{n-m}]^2 \qquad (19)$$

Moreover, two roots of $P(z)$ are known and are equal to plus and minus unity. Then we must have

$$\sum_{m=1}^{M} p_m = -1 \qquad (20.a)$$

$$\sum_{m=1}^{M} (-1)^m \, p_m = -1 \tag{20.b}$$

To minimize ε^2 with the conditions (20), we use the Lagrange multiplier method. We look for the minimum of

$$F = \varepsilon^2 - \lambda \sum_{m=1}^{M} p_m - \mu \sum_{m=1} (-1)^m \, p_m \tag{21}$$

with respect to the p_m's. The coefficients λ and μ will be calculated such as to verify (20).

Taking the partial derivatives of (21) lead to the following linear system

$$\sum_{m=1}^{M} p_m \Big[\sum_{n=M}^{N} u_{n-m} u_{n-j} \Big] = - \sum_{n=M}^{N} u_n u_{n-j} \quad \frac{\lambda}{2} \quad \frac{\mu}{2} (-1)^j \qquad j=1,\dots M \tag{22}$$

Its solution can be written

$$p_j = p_j^0 + \frac{\lambda}{2} \, p_j^1 + \frac{\mu}{2} \, p_j^2 \tag{23}$$

The different terms p_j^0, p_j^1 and p_j^2 are calculated by solving three times the linear equations system (23) with the appropriate right hand side, respectively $-\sum u_n u_{n-j}$, 1 and $(-1)^j$. The p_j's also must satisfy the conditions (20). We get a two-linear equations system

$$\frac{\lambda}{2} \sum_{j=1}^{M} p_j^1 + \frac{\mu}{2} \sum_{j=1}^{M} p_j^2 = -1 - \sum_{j=1}^{M} p_j^0 \tag{24.a}$$

$$\frac{\lambda}{2} \sum_{j=1}^{M} (-1)^j \, p_j^1 + \frac{\mu}{2} \sum_{j=1}^{M} (-1)^j \, p_j^2 = -1 - \sum_{j=1}^{M} (-1)^j \, p_j^0 \tag{24.b}$$

we solve to get λ and μ and to calculate the p_j's which minimize ε^2 and satisfy the conditions (20). The others values of r_m are then equal to the roots of P(z). All these roots must have a modulus smaller than unity to ensure the convergence of (10) at large n. Roots with a modulus larger than unity can occur if M is too large [3]. The procedure to determine M is then the following. First we choose M as large as possible, that is N/2. This is just the number of pairs of recursion coefficients a_n and b_n we want to extrapolate. If one or several roots occur with a modulus larger than unity, we decrease M by 1 until all roots have a modulus smaller than 1. Then the amplitudes C_m are also determined by a least square fit of (10). Now we can use the equation (10) to extrapolate the coefficients a_n and b_n to very large n.

Let us consider now the case where the limits a and b are known. In this case, instead of the equation (7) to define u_n, we take

$$u_{2n-1} = (a_n - a)/2 \tag{25.a}$$

$$u_{2n} \quad = b_n - b \tag{25.b}$$

No roots of the polynomial P(z) are known and its coefficients are just equal to the p_j^0. The procedure is then exactly the same to determine M and to extrapolate the coefficients at large n.

3 APPLICATION TO SILICON BAND STRUCTURE

As a test of our procedure, we have chosen to calculate the s and p electronic densities of states of bulk Si. These densities of states present all the features we want to reproduce or to improve. They can be also accurately calculated by integration over the Brillouin zone by the tetrahedron method using a linear [10] or parabolic [11] variation of the energy. We have used at most 21 exact a_n and 20 b_n (N=41) for the s and p states. The figures 1.a and 1.b show the fits we get with 20 and 15 exponentials for respectively the s and p coefficients. Fits with larger values of M give some P(z) roots with a modulus larger than unity. They have been rejected. Table 1 gives the band limits we get using the a and b

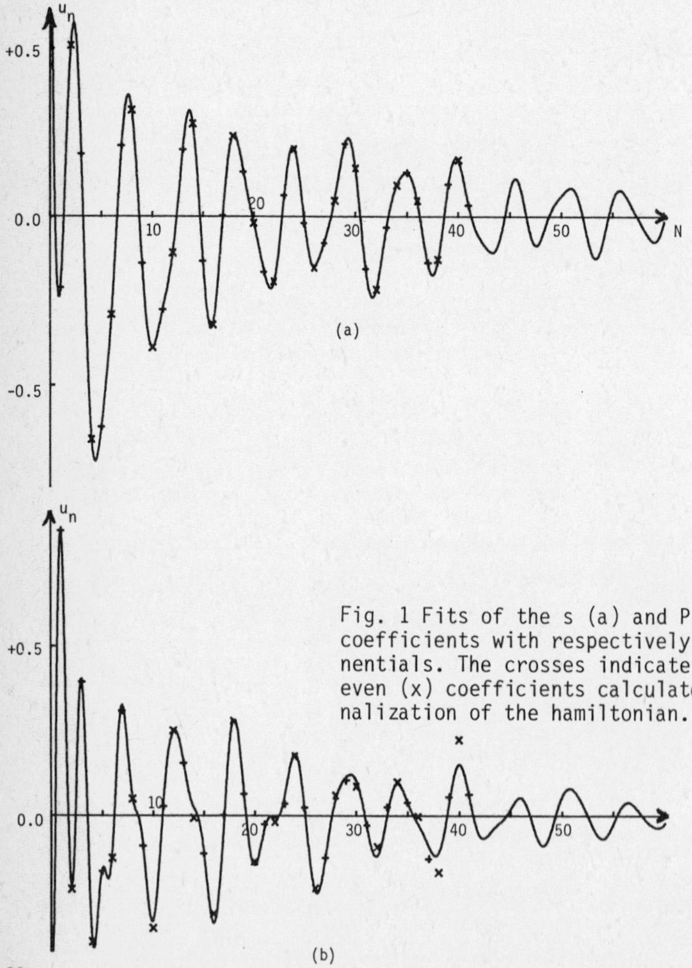

Fig. 1 Fits of the s (a) and P (b) recursion coefficients with respectively 20 and 15 exponentials. The crosses indicate the odd (+) and even (x) coefficients calculated by triadiagonalization of the hamiltonian.

Table 1 Bulk Si band limits (in eV) deduced from the recursion coefficients and compared to the exact ones. E_T (G_T) is the top (bottom) of the conduction band. E_B (G_B) is the bottom (top) of the valence band.

	s	p	exact
E_T	5.904	5.945	5.992
G_T	1.312	1.366	1.202
G_B	-0.560	-0.155	0.060
E_B	-12.214	-12.167	-12.222

calculated values. The agreement with the exact values is very good. We also compare the gap limits given by the perturbation theory to the exact ones. In fact the P(z) roots corresponding to the band gap has a modulus close to unity (0.971 and 0.968 for the s and p gaps). The gap limits we obtain are also close to the exact ones. The overall accuracy is of the order of 0.1 eV.

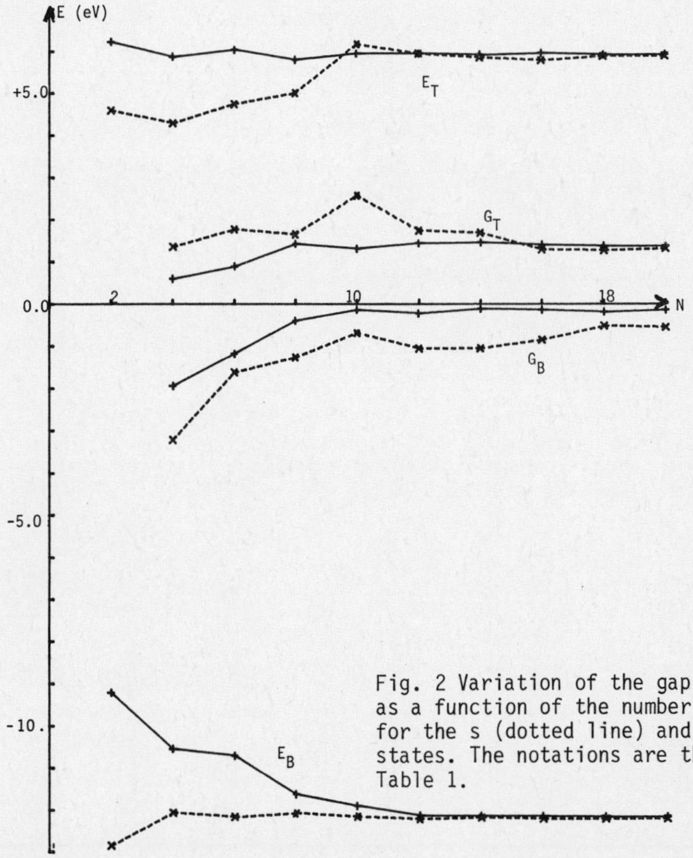

Fig. 2 Variation of the gap and band limits as a function of the number of coefficients for the s (dotted line) and p (full line) states. The notations are the same as in Table 1.

Figure 2 shows the variation of the gap and band limits as a function of the number of known coefficients. These limits are rather constant for n larger than 12. This number seems to be the limit to get a quite good density of states.

We have used the expression (10) to calculate a large number of extrapolated coefficients (between 200 and 400 so that the remaining oscillation around a and b becomes small). In spite of our almost sinusoidal termination, we have not found any spurious gap [7] in the density of states. The figure 3 shows the s, p and total density of states calculated with our procedure.

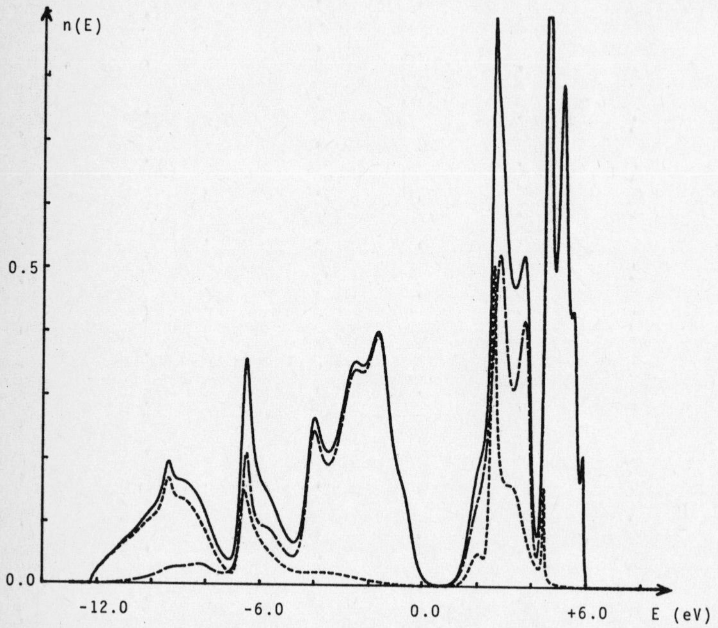

Fig. 3 Bulk Si densities of states calculated by our linear predictive model: s states (dotted line), p states (dashed dotted line), total (full line).

In figure 4, we compare the total density with an "exact" one calculated by integration over the Brillouin zone using a parabolic variation of the energy [11]. The agreement between both densities is very good.

4 CONCLUSION

We have given a simple procedure to obtain a continued fraction termination and to extrapolate the recursion coefficients [12]. In all the cases we have considered, our procedure gives very good values of the band limits which are generally not known when one uses the recursion method. These limits can then be used for a constant coefficient termination of the continued fraction. As we can extrapolate and use a large number of coefficients and of levels in the continued fraction, the matching between the constant coefficients and the extrapolated ones is very smooth. Spurious oscillations in the density of states are then avoided. It also improves the overall behaviour of the density of states and notably reduces

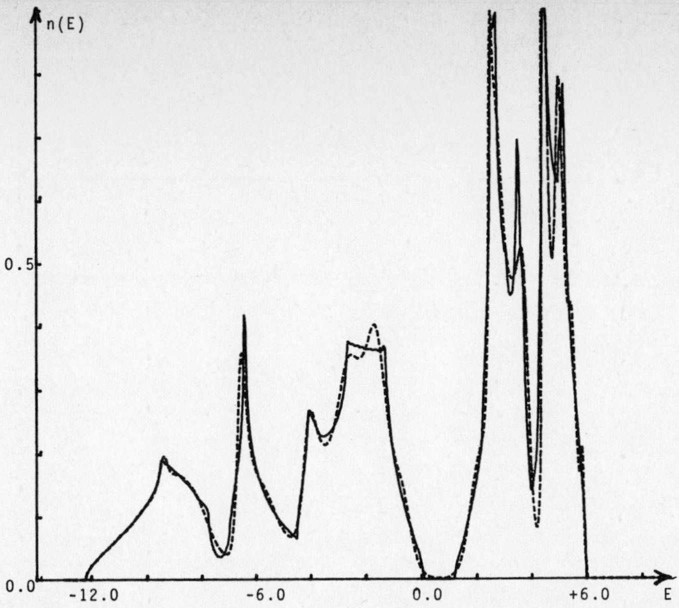

Fig. 4 Comparison of the density of states calculated with extrapolated coefficients (dotted line) and an "exact" one obtained by integration over the Brillouin zone (full line).

its value inside the gaps. In any case, it improves the convergence of the density of states obtained by the recursion method to the exact one.

REFERENCES

1. See the paper of V. Heine in this volume
2. See the papers in this volume dealing with the asymptotic behaviour of the recursion coefficients and the continued fraction termination
3. J. Makhoul, Proc IEEE 63,561(1975)
4. R. Haydock, V. Heine, M.J. Kelly, J.Phys. C 5,2845(1972)
5. J.P. Gaspard, F. Cyrot-Lackmann, J.Phys.C 6,3077(1973)
6. D.M. Bylander,J.J. Rehr, J.Phys. C 13,4157(1980)
7. P. Turchi, F. Ducastelle, G. Treglia, J.Phys.C 15,2891 (1981)
8. C.L. Hodges, J.Physique 38,187(1977)
9. A. Trias, M. Kiwi, M. Weissmann, Phys.Rev. B28,1859(1983)
10. J. Rath,A.J. Freeman,Phys.Rev. B11,2109(1975)
11. M.S. Methfessel, M.H. Boon, F.M. Mueller,J.Phys. C 16,L949 (1983)
12. G. Allan, J. Phys. C 17,3945 (1984)

Part III

Related Methods

On a Generalized-Moments Method

J.P. Gaspard

Université de Liège, Institut de Physique B5, B-4000 Sart Tilman, Belgium

Ph. Lambin

FUNDP, Département de Physique, 61 rue de Bruxelles, B-5000 Namur, Belgium

The continued-fraction technique has been widely used in solid state physics. It provides a useful theoretical framework for calculating the densities of states (of electrons, phonons...) mainly in disordered systems. The continued-fraction coefficients are computed either from the moments of the density of states or more directly by the recursion method. The former has the advantage of the linearity upon the density of states and the latter is numerically optimized. The generalized-moments method is an interpolation between both methods that, if suitably used, combines their advantages. A version of the method could be seen as a perturbation expansion from the Bethe lattice. The asymptotic limits of the continued-fraction coefficients are shown to be accurately determined from the generalized-moments associated to the recursion method. Computations of electronic densities of states illustrate the method.

1. Introduction

There is a large number of physical situations where a positive definite function (e.g. density of excitations) has to be computed from the knowledge of a small number of its moments. Typically, this situation is often encountered in physics of non-crystalline solids where the usual techniques of band-structure calculation (mainly based on the Bloch theorem) are no longer applicable, whereas the moments of the density of excitations (electrons, phonons,...) can still be computed. For such applications, it is important to develop mathematical methods that are close to the physics involved. Hence the importance of direct-space methods that are directly relevant to non-perfect crystals. These direct-space methods have been widely used, since more than fifteen years, in the field of non-crystalline materials involving surfaces, amorphous systems and liquids, alloys,and also point defects (impurities, vacancies, etc...)or dislocations. Of paramount importance when investigating properties of non-periodic systems is the local density of states which reflects the spatial variations of the electronic (or vibrational, magnetic,...) properties. Precisely, such local density of states is directly accessible to real-space calculations. The technique used to characterize (often roughly) the (local) density of excitations, n(E) say, directly recurs to the famous "moment problem" which dates back to the last century (Tchebytcheff, Stieltjes [1],...).This theorem states that, under not very restrictive conditions, n(E) is determined by the complete set of its moments [2].
In practice, n(E) is computed by the way of its Hilbert transform R(z), using the formula

$$n(E) = -\frac{1}{\pi} \lim_{\varepsilon \to 0} \operatorname{Im} R(E+i\varepsilon), \tag{1}$$

where R(z) is written as the continued fraction

$$R(z) = \cfrac{b_o}{z-a_1 - \cfrac{b_1}{z-a_2 - \cfrac{b_2}{z-a_3 - \ddots}}} \tag{2}$$

It can be shown that the coefficients a_i, b_i of the continued fraction can be computed from the moments of n(E). Unfortunately, the calculation of the continued-fraction coefficients from the moments is notoriously an ill-conditioned numerical algorithm (see below).

The now classical "recursion method" [3] allows a direct calculation of the continued fraction-coefficients $\{a_i, b_i\}$ from the Hamiltonian, corresponding to the local density of states at a given site (and a given orbital) of the structure. This method is numerically optimized as shown in section 3, but the calculated quantities are no longer linear upon n(E). By contrast, the moments are linear functionals of n(E). This is particularly important for disordered systems, when configurational average of the density of states has to be performed. While the moments can be averaged, the a_i and b_i coefficients cannot. The non-linearity constitutes a pitfall of the recursion method. In addition, the continued-fraction coefficients do not authorize any straightforward derivation of a convolution of n(E), while the moments do.

As we show in this paper, the generalized-moments method [4] combines the advantages of the power moments method (linearity) and the recursion technique (stability). Moreover, this method is an interesting starting point to perform perturbation calculations. However, the numerical stability of the generalized-moment method is conditioned by the choice of the parameters as shown in section 4.

2. The power moments

The power moments of a function n(E) are defined as

$$\mu_k = \int_{-\infty}^{+\infty} E^k \, n(E)dE \quad , \quad k=0,1,\ldots \quad . \tag{3}$$

In a tight binding description of one-body excitations, where the Hamiltonian H is given in an orthogonal basis localized on the atomic sites and orbitals of the system, the moment μ_k of n(E) can be expressed as follows :

$$\mu_k = \langle 0|H^k|0\rangle = \sum_{i_1\ldots i_{k-1}} H_{0i_1} H_{i_1 i_2} \ldots H_{i_{k-1}0} \quad . \tag{4}$$

In this expression, 0 denotes the starting element of the basis and i_j are intermediate elements, while H_{ij} are the components of the Hamiltonian in the local basis. One interesting aspect of the power moments is their direct link to the structure of the lattice so that the relations between the atomic structure and the density of excitations can be found.

In the case of the electronic eigenstates, for s states (without orbital degeneracy) when the diagonal elements of the Hamiltonian are zero and if the resonance integrals between neighbouring sites are all identical to β, formula (4) yields

$$\mu_k = \beta^k \Sigma S_{0i_1} S_{i_1 i_2} \ldots S_{i_{k-1} 0} \qquad (4')$$

where S_{ij} is the connectivity matrix, i.e. $S_{ij}=1$ when i and j are neighbouring sites and 0 otherwise. Consequently the moment of order k is the sum of contributions of all the closed walks of length k that can be made along the connections of the structure, starting from the site 0. For instance, in a simple cubic lattice with nearest-neighbour interactions equal to β, the second moment is equal to $6\beta^2$ where 6 is the coordination. As there are no triangular circuits connecting nearest neighbours, the third moment is zero. The fourth moment writes, pictorially,

$\mu_4 =$ 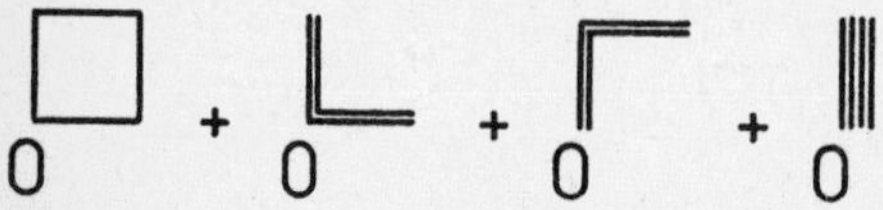 $= 24\beta^4 + 30\beta^4 + 30\beta^4 + 6\beta^4 = 90\beta^4$.

The formulas become more complex in presence of orbital degeneracy; however the first moments can still be calculated by hand. In fact, calculation of the moments of higher order is easy to perform on a computer, as they are powers of the Hamiltonian H. If we keep the resonance integrals of the σ bonds only (i.e. the sole ppσ contribution for the p bands, the sole ddσ contribution for the d bands ,[5]...) formula (4') gives for p bands

$$\mu_k = \beta^k \Sigma S_{0i_1} S_{i_1 i_2} \ldots S_{i_{k-1}0} \cos\hat{0} \cos\hat{i}_1 \ldots \cos\hat{i}_{k-1} \qquad (4'')$$

and more generally

$$\mu_k = \beta^k \Sigma S_{0i_1} S_{i_1 i_2} \ldots S_{i_{k-1} 0} P_\ell(\cos\hat{0}) P_\ell(\cos\hat{i}_1) - P_\ell(\cos\hat{i}_{n-1}) \qquad (4''')$$

where β is the σ bond resonance integral, $\hat{i}_j$ is the angle of the $i_{j-1}i_j$ and $i_j i_{j+1}$ bonds and P_ℓ is the Legendre polynomial of the order ℓ of the degeneracy (0 for the s band, 1 for p bands...). In the case of the p bands on a simple cubic, only the straight line contributions are kept, so that μ_4 reduces to $18\beta^2$, identical to the value related to three linear chains along x, y and z.

From the first few moments, it is possible to characterize grossly the density of states. By a normalization to unity and a suitable choice of the zero of energy, it is always possible to take $\mu_0=1$ and $\mu_1=0$. In these conditions, the second moment of the density of states μ_2 characterizes the effective width w of the spectrum :

$$w = a\sqrt{\mu_2} \qquad (5)$$

where a is a parameter dependent weakly on the shape on n(E) and close to 3 (it is equal to $2\sqrt{3}=3.46$ for a rectangular n(E), equal to $2\sqrt{2\ell n2}=2.35$ for the FWMH of a gaussian curve and $\sqrt{6\pi}=4.34$ for the semi-elliptic band).

The dimensionless quantity $\mu_3/\mu_2^{3/2}$ is sensitive to the asymmetry of the band (as are the odd order moments); it is zero for a symmetric band (the exact condition for having a symmetric band is that the odd order moments vanish). From (4") and (4"') we observe that in structures containing triangular circuits the orbital degeneracy ℓ decreases substantially the value of $\mu_3/\mu_2^{3/2}$ so that the density of states becomes more symmetric by increasing ℓ^2 (the angular term of an equilateral triangle is 1(ℓ=0), 1/8(ℓ=1), -1/512(ℓ=2)...).

The fourth order moment gives information on the overall shape of the band. More precisely, the dimensionless quantity s

$$s = \frac{b_2}{b_1} = \frac{\mu_4\mu_2 - \mu_2^3 - \mu_3^2}{\mu_2^3} \qquad (6)$$

discriminates between the densities of states with a central peak (unimodal distribution of the states), s>2, and the bimodal distributions (two well separated peaks), s<1. The intermediate region 1<s<2 corresponds to shapes of n(E) close to a rectangular density (s=0,8) (Fig.1). As we will see later, b_1 and b_2 in equation (6) are the coefficients of a continued fraction.

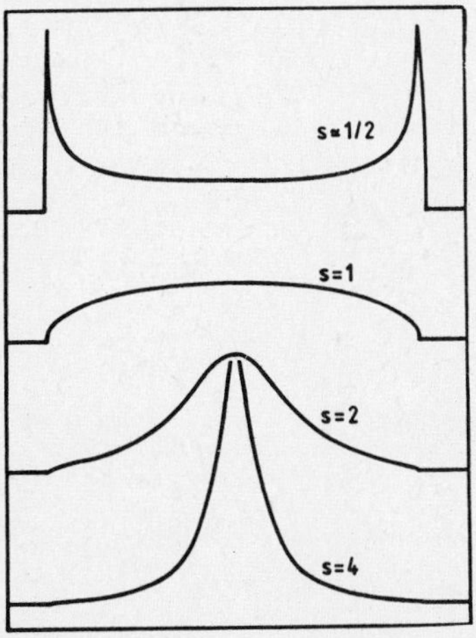

Fig. 1 Densities of states computed for several values of the parameter s (Eq. 6), showing the transition between bimodal distribution (top of the drawing) to unimodal distribution (bottom).

Beyond this rough characterization of density of states by a limited number of moments, a larger number of moments is required to construct n(E). Roughly, one can estimate that the resolution Γ of the density of states is

$$\Gamma = \frac{W}{N}$$

where W is the bandwidth and N the number of moments. Of course, the exact calculation of n(E) requires an infinite number of moments. In simple topologies , and s-band Hamiltonian, the moments can be obtained analytically (e.g. linear chain, Bethe lattice...) or by series expansions (simple cubic, fcc...) [6].

The calculation of the moments of a given Hamiltonian H on a computer is, in principle, straightforward. Practically, the moments are calculated by the formulas

$$\mu_{2i} = \sum_k \langle 0|H^i|k\rangle\langle k|H^i|0\rangle \qquad (7)$$

$$\mu_{2i-1} = \sum_k \langle 0|H^{i-1}|k\rangle\langle k|H^i|0\rangle \quad . \tag{7'}$$

These expressions need to store at the same time the two vectors $\langle k|H^{i-1}|0\rangle$ and $\langle k|H^i|0\rangle$, rather than the matrices H^{i-1} and H^i. However, the size of the vector $\langle k|H^i|0\rangle$ increases as the third power of i in three-dimensional materials. The storage of $\langle k|H^{i-1}|0\rangle$ and $\langle k|H^i|0\rangle$ and the number of multiplications involved when constructing recursively these vectors constitute the practical limitation on the number of moments accessible.

The coefficients of the continued fraction (2) can be obtained from the moments by the following formulae

$$b_i = \frac{\Delta_i \Delta_{i-2}}{\Delta_{i-1}^2} \quad \text{and} \tag{8}$$

$$a_i = \frac{1}{\Delta'_{i-2}} \left(\frac{\Delta_{i-1}\Delta'_{i-3}}{\Delta_{i-2}} + \frac{\Delta_{i-2}\Delta'_{i-1}}{\Delta_{i-1}} \right) \tag{8'}$$

where Δ_i is the Hankel determinant

$$\Delta_i = \det \begin{bmatrix} \mu_0 & \mu_1 & \cdots & \mu_i \\ \mu_1 & \mu_2 & \cdots & \mu_{i+1} \\ & \cdots\cdots & & \\ \mu_i & \mu_{i+1} & & \mu_{2i} \end{bmatrix} \quad \text{and} \tag{9}$$

$$\Delta'_i = \det \begin{bmatrix} \mu_1 & \mu_2 & \cdots & \mu_{i+1} \\ \mu_2 & \mu_3 & \cdots & \mu_{i+2} \\ & & \cdots & \\ \mu_{i+1} & \mu_{i+2} & \cdots & \mu_{2i+1} \end{bmatrix} \quad . \tag{9'}$$

Formulae (9) and (9') are notoriously ill-conditioned : the determinants Δ_i (resp Δ'_i) become increasingly small [relatively to the terms $\mu_0\mu_2\cdots\mu_{2i}$ (resp $\mu_1\cdots\mu_{2i+1}$)].

Practically, in double precision arithmetics (with 64 bits per word), formula (9) or (9') gives no more significant digit for i>15; consequently a larger number of continued-fraction coefficients requires multiple-precision arithmetics. This is due to the fact that the moments store inefficiently the information. As we see in section 3 and 4 the recursion method and the generalized-moments method circumvent this effect. Incidentally, formulae (9) and (9') show that the continued-fraction coefficients a_i and b_i are strongly non-linear functions of the moments.

3. The recursion method

The recursion method is now classically described in a series of review papers [3]. It closely resembles the Lanczos [7] tridiagonalization of hermitian matrices. The idea is to transform the Hamiltonian matrix into a tridiagonal matrix by a series of unitary transformations. On a more geometric formulation, the unitary transformations convert the Hamiltonian on the real lattice into an equivalent Hamiltonian on a semi-infinite chain H_{TD} (Fig. 2), with

$$H_{TD} = \begin{pmatrix} a_1 & \sqrt{b_1} & 0 & \\ \sqrt{b_1} & a_2 & \sqrt{b_2} & \\ 0 & \ddots & \ddots & \ddots \end{pmatrix} . \tag{10}$$

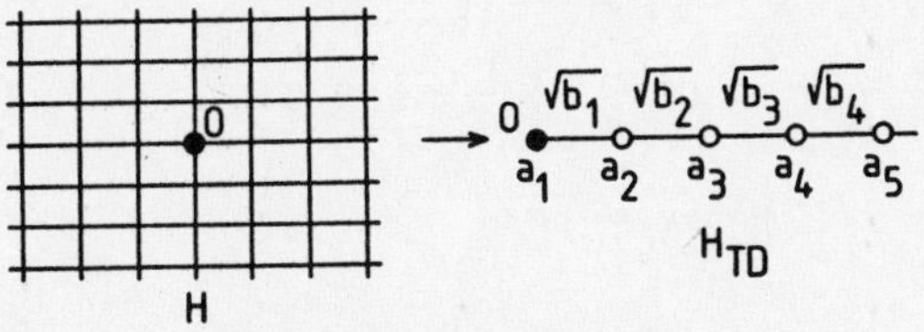

Fig. 2 The recursion method maps the real structure onto a semi-infinite chain of atoms.

The coefficients a_i are the energy levels and $\sqrt{b_i}$ are the nearest-neighbour hopping (or resonance) integrals. Notice that the interpretation of the coefficients a_i and b_i in terms of the lattice structure is not straightforward (by contrast to power moments). We do not dwell any longer on the description of the recursion method as it appears as a special (self-consistent) case of the generalized-moments method. One of the advantages of the recursion method is its numerical stability (cf.section 4). However it is not possible to calculate directly averaged densities of states with the recursion method.

4. The generalized-moments method

The underlying idea of the method is to substract from the power moments the non relevant-contributions coming from lower order moments, so that the transformations modified moments → $\{a_i, b_i\}$ is well conditioned. This idea has been developed by several authors [8,9] who introduced the modified moments m_k

$$m_k = \int_{-\infty}^{+\infty} P_k(E)\, n(E) dE \tag{11}$$

where P_k is a k-th order polynomial of E.

This concept was implemented in physics [10] for the computation of frequency distributions of harmonic solids. The choice of the polynomials P_k is, of course, determinant in the stability of the calculation of the a_i, b_i coefficients: the closer the P's to the orthogonal polynomials, the more stable the algorithm. It proceeds as follows : from a suitable choice of the P_k's, it is possible to compute the Gram matrix

$$G_{ij} = \int_{-\infty}^{+\infty} P_i(E)\, P_j(E) n(E) dE \qquad \begin{matrix} 0 \leq i \leq k \\ 0 \leq j \leq k \end{matrix} \tag{12}$$

from the modified moments only. This requires,of course,computing the modified moments up to order 2k. This being achieved, the Gram matrix is transformed into a tridiagonal form using e.g. the Lanczos method leading to the coefficients a_i, b_i $(i=1...k)$. Of course,the numerical stability of the algorithm relies upon the direct calculation of the m_k's from the Hamiltonian. The calculation of the modified moments from the power moments (expanding P_k in power series) would be inefficient. Let us also emphasize that the computations of $\{a_i, b_i\}_{i=1,.,k}$ requires the computation of the

set $\{m_0, m_1, .., m_{2k}\}$. That is to say, powers of H up to the order 2k. By contrast, as shown below, the generalized-moments method requires the calculation of powers of H up to order k only.

The generalized-moments ν_{2i} and ν_{2i-1} are defined as follows :

$$\nu_{2i} = G_{ii} = \int_{-\infty}^{+\infty} P_i(E)\, P_i(E)\, n(E)\, dE \tag{13}$$

$$\nu_{2i-1} = G_{i-1,i} = \int_{-\infty}^{+\infty} P_{i-1}(E)\, P_i(E)\, n(E)\, dE \quad . \tag{13'}$$

In addition, the polynomials $P_i(E)$ must satisfy a three-term recurrence equations :

$$P_{i+1}(E) = (E - c_{i+1})\, P_i(E) - d_i\, P_{i-1}(E) \tag{14}$$

with the initial conditions

$$P_{-1}(E) = 0$$

$$P_0\,(E) = 1 \,.$$

Indeed, equation 14 allows to fill in the Gram matrix recursively from its diagonal and subdiagonal elements.

Besides, the generalized-moments are obtained in a way as simple as the power moments

$$\nu_{2i} = \sum_j \langle 0|\, P_i(H)\,|j\rangle\langle j|\, P_i(H)\,|0\rangle \tag{15}$$

$$\nu_{2i-1} = \sum_j \langle 0|\, P_{i-1}(H)\,|j\rangle\langle j|\, P_i(H)\,|0\rangle \tag{15'}$$

which is the generalization of formulae (7) and (7'). The vectors $\langle 0|P_i(H)|j\rangle$ $(j=0,...N)$ are computed recursively from the Hamiltonian by

$$\begin{aligned}\langle 0|P_{i+1}(H)|j\rangle = \sum_m \langle 0|H|m\rangle\langle m|\, P_i(H)\,|j\rangle \\ - c_{i+1}\, \langle 0|\, P_i(H)\,|j\rangle \\ - d_i\, \langle 0|\, P_{i-1}(H)\,|j\rangle \;.\end{aligned} \tag{16}$$

The computation of the vectors follows exactly the recursion algorithm , except that the coefficients c_i and d_i are defined a priori. In the recursion method, however, they are calculated self-consistently in order to orthogonalize the polynomials with respect to n(E) (consequently the

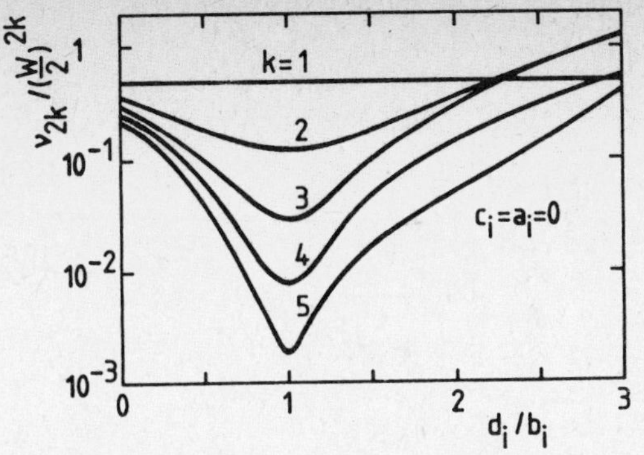

Fig. 3 Generalized moments of an infinite chain of atoms computed from the polynomials sequence defined by Eq. 14. The even-order generalized moments are minimum when the polynomials are orthogonal with respect to the density of states, i.e. $c_i=a_i$, $d_i=b_i$ (recursion method). The power moments correspond to $c_i=0$, $d_i=0$.

recursion vectors are orthogonal). On the other hand, the power moments correspond to $c_i=0$ and $d_i=0$. We should stress that the generalized moments are still linear functionals of n(E), so that they can be averaged in the same way as the power moments.

The numerical stability of the calculation of the a_i's and b_i's relies upon the choice of the c_i's and d_i's. We have observed that a choice of constant values of c_i and d_i consistent with the band limits (e.g. $c_i=m$, middle of the band, and $d_i=W^2/16$, W is the bandwidth) gives a numerical stability close to the recursion method, as illustrated in Fig. 3 . The minimum of all the curves in Fig. 3 are obtained for $c_i=a_i$ and $d_i=b_i$, e.g. the recursion algorithm. In fact it is easy to show that the generalized-moments of even order are minimal over the all set of monic polynomials $P_k(E)$ precisely when $P_k(E)$ is orthogonal with respect to n(E) [4], in which case

$$\nu_{2i} = b_1 \ldots . b_i \quad \text{and} \tag{17}$$

$$\nu_{2i-1} = b_1 \ldots . b_{i-1}(a_1+\ldots+a_i). \tag{17'}$$

In the simplest case of a Bethe lattice, with constant hopping elements, the minimal moments can be interpreted easily in terms of walks on the lattice. They correspond to the self-retracing walks. In other words, the minimals moments ν_{2i} contains only the contribution coming from the walks that reach the ith shell of neighbours. This is strictly no longer true on other lattices : the redundant information is still removed "as best as possible" in the recursion method.

The computation time is, within a few percents, identical for the power-moments method, the recursion method and the generalized-moments method.

5. Perturbative aspects of the Generalized-moments method

It often occurs that two physical systems differ by a small perturbation of their Hamiltonian H (e.g. a weak distortion of a perfect crystal by phonons). In that case,it is not without interest to deduce the density of states n(E) of the perturbed system from the density of states $n_0(E)$ of the reference system by a perturbation technique. In the present section, we show that the generalized-moments method is particularly well designed for perturbation expansions. We assume that the continued fraction coefficients c_k and d_k of the unperturbed density of states $n_0(E)$ are known, from which the orthogonal polynomials $P_k(E)$ can be constructed. Let us write [10]

$$n(E) = n_0(E) + \sum_{k=1}^{\infty} \alpha_k P_k(E) n_0(E) \quad . \tag{18}$$

We assume that $n_0(E)$ and $n(E)$ have identical band limits in order to avoid Gibbs oscillations near the band edges. Also $n_0(E)$ must be "close" to $n(E)$ in order to fasten the convergence of the series. Since the polynomials $P_k(E)$ are orthogonal with respect to $n_0(E)$, the coefficients α_k of the series are obviously given by

$$\alpha_k = \frac{\int P_k(E) n(E) dE}{\int P_k^2(E) n_0(E) dE} \quad , \qquad k=1,2,\ldots \quad . \tag{19}$$

The denominator of this expression reduces to the product $d_1 \ldots d_k$. The numerator is the kth order modified moment m_k of $n(E)$ (Eq. 11). Consequently,

$$\alpha_k = \frac{m_k}{d_1 \ldots d_k} \quad , \qquad k=1,2,\ldots \quad . \tag{20}$$

Besides, m_k is easily computed recursively from Eq. 16 and from

$$m_k = \langle 0 | \, P_k(H) | 0 \rangle \tag{21}$$

where $|0\rangle$ is the site (including the orbital) where the local density of states of the perturbed Hamiltonian is computed.

As an illustrative application of this technique, we consider a semi-infinite linear chain with first and second nearest-neighbour interactions (the corresponding tight binding parameters are β and γ respectively). For a Hamiltonian with only nearest-neighbour interactions the band extends from $-2|\beta|$ to $2|\beta|$. When $|\gamma/\beta|<1/4$, the bandwidth is unchanged but the band is shifted by 2γ (i.e. it extends from $-2|\beta|+2\gamma$ to $2|\beta|+2\gamma$). In order to calculate the local density of states on the "surface" of the semi-infinite chain, let us take the following set of continued-fraction coefficients

$$c_k = 2\gamma \quad \text{and} \quad d_k = \beta^2 \qquad k=1,2,\ldots \tag{22}$$

The related unperturbed density of states

$$n_0(E) = \frac{\sqrt{4\beta^2-(E-2\gamma)^2}}{2\pi\beta^2} \qquad |E-2\gamma|<2|\beta| \tag{23}$$

corresponds to a semi-infinite linear chain with first neighbour interactions β only and atomic levels 2γ. In these conditions, the modified moments of $n(E)$ are readily obtained as

$$m_k = (-\gamma)^k \, \frac{(2k+2)!}{(k+1)!(k+2)!} \quad . \tag{24}$$

Finally, the $P_k(E)$ are the shifted Tchebytcheff polynomials of the second kind U_k and $n(E)$ writes

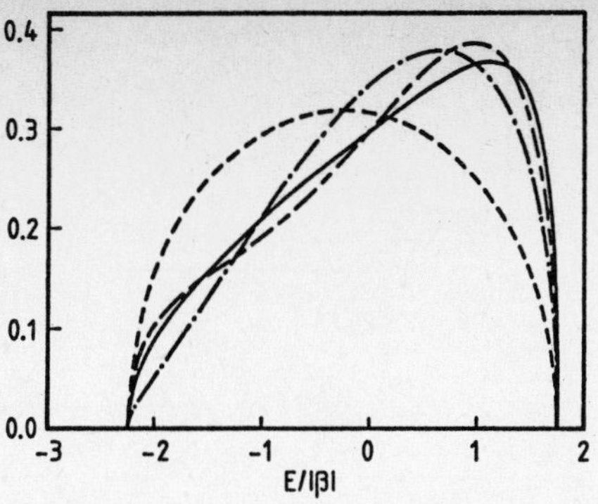

Fig. 4 Convergence of the perturbation series (25). The dashed curve shows the unperturbed density of states $n_0(E)$ (Eq. 23). The full curve gives the perturbed distribution $n(E)$. The other curves correspond to the first-order (—·—·—) and second order (— — — —) perturbed series. These calculations were performed for $|\beta/\gamma| = 1/8$.

$$n(E) = n_0(E)\left[1 + \sum_{k=1}^{\infty} \frac{(2k+2)!}{(k+1)!(k+2)!} \left(-\frac{\gamma}{|\beta|}\right)^k U_k\left(\frac{E-2\gamma}{2|\beta|}\right)\right] \quad . \qquad (25)$$

The series is shown to be convergent for $|\gamma/\beta|<1/4$. Typical results are shown in Fig.4 for $|\gamma/\beta|=1/8$.

6. Band edge singularities and asymptotic behaviour of a_i and b_i

In the cases where the band edges are not known a priori, they are determined by extrapolation of the coefficients a_i and b_i. We show in this section that it is preferable to extrapolate the minimum moments (=the generalized moments corresponding to the recursion coefficients) than the a_i and b_i themselves. For a one-band density of states, the coefficients a_i and b_i have the well-known asymptotic limits

$$a_\infty = (E_u+E_l)/2$$
$$b_\infty = (E_u-E_l)^2/16 \qquad (26)$$

where E_u and E_l are the upper and lower edges. However, the coefficients are usually scattered around the asymptotic values(due to features of the local densities of states or due to oscillations produced by internal singularities), so that the asymptotic limits cannot be accurately determined. The consideration of the minimum generalized-moments ν_k allows to determine the asymptotic limits and the band edge singularities with a greater accuracy. Let us consider

$$\nu_{2k} = b_1 b_2 \ldots b_k \quad . \qquad (27)$$

It is obvious that this quantity (or its kth root) is more stable than the individual b_k's. Moreover, while the correction in the b_k's from their asymptotic behavior is of the order k^{-2} (due to band edge singularities), it is of the order k^{-1} in the ν_{2k}. More specifically, if the band edge behavior is

$$n(E) \sim (E_u-E)^\alpha (E-E_l)^\beta, \quad \alpha,\beta > -1 \qquad (28)$$

- a common occurence in solid state physics- the minimum moments behave as

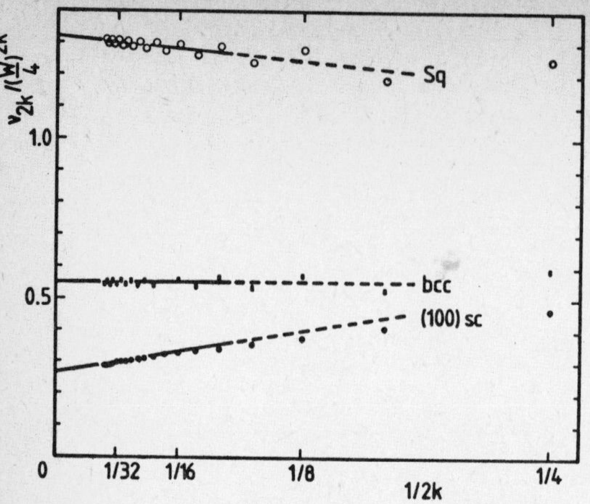

Fig. 5 Plots of normalized minimum moments (s band) for two-dimensional (square lattice) and three-dimensional (bcc) systems and for the (100) surface of the simple cubic lattice.

$$\nu_{2k} \sim c\left(1+\frac{\alpha^2+\beta^2-.5}{2k}\right)\left[(E_u - E_l)/4\right]^{2k}, \quad k \to \infty . \tag{29}$$

The slope of plots of $\nu_{2k}/(W/4)^{2k}$ vs k^{-1} is proportional to $\alpha^2+\beta^2-.5$ which is equal to $-.5$ for two-dimensional systems (α=β=0), to zero for three-dimensional systems (α=β=1/2) and to 4 for surfaces (α=β=3/2). Fig.5 illustrates these behaviours. A similar analysis can be made on the a_k coefficients [4].

Appendix A - Moments and convolution

It is sometimes interesting to convolute a density of states n(E) for which the continued-fraction coefficients are difficult to extrapolate. The power moments $\tilde{u}_k$ of the convoluted density $\tilde{n}(E)$

$$\tilde{n}(E) = n(E) * r(E) \tag{A1}$$

can be determined directly if the power moments μ_i and ν_i of n(E) and r(E) are known. Indeed, it is easy to show that

$$\tilde{\mu}_k = \sum_{i=0}^{k} \mu_i \, \nu_{k-i} \qquad k=0,1,2,... \tag{A2}$$

The convolution by a function r(E) decreases the oscillations of the coefficients $\{a_i, b_i\}$ of the continued fraction as illustrated in Fig.6.

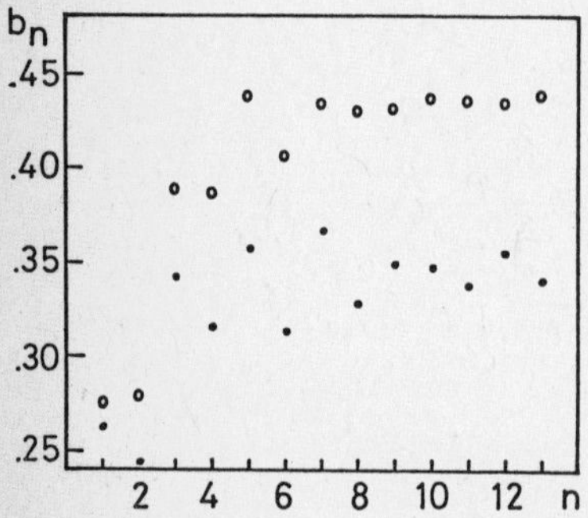

Fig. 6 Continued-fraction coefficients b_n : bare values (•) and after convolution (o) by a function r(E) the width of which is equal to one-third of the bandwidth of the convoluted density of states.

Extension to the case of generalized moments is difficult, however not impossible. It requires to know some algebra on the orthogonal polynomials. Let us put the problem in a slightly different way using the modified moments. The monic polynomials $P_k(E)$, $P_k^{(r)}(E)$ and $\tilde{P}_k(E)$ define modified moments (the polynomials are not necessarily orthogonal) relatively to $n(E)$ $r(E)$ and $\tilde{n}(E)$ respectively.
It is always possible to express $\tilde{P}_k(\alpha+\beta)$ as a sum of products of $P_k(\alpha)$ and $P_k^{(r)}(\beta)$, generalizing the binomial formula for $(\alpha+\beta)^k$, i.e.

$$\tilde{P}_k(\alpha+\beta) = \sum_{\substack{i=0\\j=0}}^{k} c_{ij}^{(k)} P_i(\alpha) P_j^{(r)}(\beta) \ . \tag{A3}$$

The c_{ij} coefficients can be determined by standard algebraic calculations. Then, the modified moments ($\nu_k, \nu_k^{(r)}$, $\tilde{\nu}_k$ respectively) are connected by

$$\tilde{\nu}_k = \sum_{\substack{i=0\\j=0}}^{k} c_{ij}^{(k)} \nu_i \nu_j^{(r)} \ . \tag{A4}$$

If the P's are the monomials, and so the ν's are the ordinary power moments, $c_{ij}^{(k)}$ vanishes if $i+j$ is not equal to k.

References:

[1] T.J. Stieltjes, Ann. Fac. Sci. Univ. Toulouse, 8, 93 (1895)
[2] N.I. Akhiezer : The Classical Moment Problem (Hafner Publ. Co., N.Y.1965)
[3] R.Haydock in Solid State Physics, ed. by H.Ehrenreich, F.Seitz, and D. Turnbull (Academic, N.Y. 35, 215, 1980)
[4] P. Lambin and J.P. Gaspard, Phys. Rev. B 26, 4356 (1982)
[5] J.C.Slater and G.F.Koster, Phys. Rev. 94, 1498 (1954)
[6] C. Domb, Ann. Phys.28, 531 (1960)
[7] C. Lanczos, J. Res. Nat. Bur. Stand. 45, 255 (1950)
[8] W. Gautschi, Math. Comp. 22, 252 (1968); ibid 24, 245 (1970)
[9] J.C.Wheeler and C. Blumstein, Phys. Rev. B 6, 4380 (1972)
[10] J.C.Wheeler, M.G.Prais and C.Blumstein, Phys. Rev. B 10, 2429 (1974)

The Equation of Motion Method

A. MacKinnon,

The Blackett Laboratory, Imperial College, Prince Consort Road,
London SW7 2BZ, United Kingdom

The equation of motion method is introduced and compared with the recursion method. It is concluded that it can probably be made *more* efficient than Lanczos/recursion for the calculation of densities of states.

1. Introduction

It is important that a specialist conference like that on the recursion method should also consider alternative methods for the calculation of those quantities for which the recursion method has proven so successful.

This contribution will discuss the equation of motion method which was originally introduced by *Alben et al.* [1] and was later generalised to the calculation of localisation and transport quantities [2,3]. Here we shall concentrate on the calculation of densities of states.

Our conclusions contrast sharply with those of previous authors [4] who have compared the equation of motion and recursion methods. This difference depends crucially on the detailed implementation of the algorithm which will be discussed in section (3), and in particular on the use of the *leap-frog* method for solving differential equations.

2. Analytical Outline

Consider a wave function $\psi(t)$ which obeys the time-dependent Schrödinger equation

$$\frac{d\psi}{dt} = \mathcal{H}\psi \tag{1}$$

where we have chosen units such that Planck's constant, $\hbar$, is unity. Formally we can write the solution of (1) in the form

$$\psi(t) = \exp(-i\mathcal{H}t)\psi(0). \tag{2}$$

The retarded Green's function $g(t)$ may be written

$$g(t) = -i\int \psi^*(0)\psi(t)\,dr^3 \tag{3}$$

and its *Laplace* transform

$$G(Z) = \int_0^\infty \exp(iZt) g(t)\, dt \tag{4a}$$

$$= \int \psi^*(0)[Z - H]^{-1}\psi(0)\, dr^3 \tag{4b}$$

where $Z = E + i\gamma$ $(\gamma < 0)$. The local density of states on $\psi(0)$ is simply related to this by

$$\rho(E) = -\frac{1}{\pi} \operatorname{Im} G(Z). \tag{5}$$

The above definitions provide us with a means of calculating $\rho(E)$ for any $\psi(0)$:

(a) integrate (1) using $\psi(0)$ as the initial condition;

(b) calculate $g(t)$ using (3);

(c) Laplace transform $g(t)$ to obtain $G(Z)$ and $\rho(E)$.

Thus $\psi(0)$ plays a similar role to the initial state of the recursion method [5] and may be freely chosen in the most appropriate way for the problem in hand.

Typical examples could be:

(a) a particular basis function of a tight-binding model;

(b) a plane wave of wave vector **k** to obtain the spectral function $A(\mathbf{k},E)$;

(c) a random $\psi(0)$ to calculate the global density of states.

3. Numerical Implementation

(a) General

When the procedure is implemented on a computer there are two main sources of error:

i) discrete time interval δt in the integration of (1);

ii) finite integration time in (4).

In both cases the error depends to some extent on the details of the integration method used. In this discussion we shall consider only the simplest practical procedures.

(b) Time Integration

There is a considerable literature on the numerical integration of differential equations. The simplest numerically stable routine for oscillatory equations such as (1) uses the so-called *leap-frog* method [6]:

$$\psi_{n+1} = \psi_{n-1} - 2i\, \delta t\, \mathcal{H} \psi_n. \tag{6}$$

This can be rewritten in the matrix form:

$$\begin{bmatrix} \psi_{n+1} \\ \psi_n \end{bmatrix} = \begin{bmatrix} -2i\,\delta t\, \mathcal{H} & I \\ I & 0 \end{bmatrix} \begin{bmatrix} \psi_n \\ \psi_{n-1} \end{bmatrix} = \lambda \begin{bmatrix} \psi_n \\ \psi_{n-1} \end{bmatrix} \tag{7}$$

for which the eigenvalues, λ, satisfy

$$\lambda_i = -i\,\delta t\, E_i \pm \sqrt{1-\delta t^2 E_i^2} \tag{8}$$

where the E_i are the eigenvalues of $\mathcal{H}$.

For $\delta t\, E_i < 1$, $|\lambda_i|^2 = 1$ and the solutions of the difference equation are purely oscillatory, as they should be. Thus δt must be chosen such that

$$\delta t \leq |E_{max}|^{-1} \tag{9}$$

where $|E_{max}|$ is the absolute maximum value of E_i.

The frequency E^* of the oscillations of the difference equation is related to E of the differential equation by:

$$\lambda = \exp(-iE^*\delta t) \tag{10}$$

so that

$$E = \frac{\sin(E^*\delta t)}{\delta t}. \tag{11}$$

According to (11) the frequencies E^* are shifted from E for finite δt. This does *not* imply that δt should be chosen very small so that $E^* = E$, but rather that the calculated density of states, $\rho^{\delta t}(E^*)$, must be corrected for the shift using

$$\rho(E) = \rho^{\delta t}\left\{\frac{\sin^{-1}(E\,\delta t)}{\delta t}\right\}/\sqrt{1-E^2\delta t^2} \tag{12}$$

A useful trick is to choose the values of E at which $\rho(E)$ is required and then to calculate the corresponding E^*'s or λ's from (11).

(c) Laplace Transformation

Consider the infinite sum

$$S(Z) = \sum_{n=0}^{\infty} \exp(iZ\,\delta t)\psi_n\psi_0^*. \tag{13}$$

For a single component, λ, we can write

$$S(Z) = \sum_{n=0}^{\infty} \lambda^n \exp(iZ\,n\delta t) \tag{14a}$$

$$= \{1 - \exp[i(Z - E^*)\delta t]\}^{-1} \tag{14b}$$

which has a pole of weight $i/\delta t$ for $(Z - E_i^*)\delta t = 2\pi m$ where m is any integer. From this we can write

$$\rho^{\delta t}(E^*) = \left(\frac{\delta t}{\pi}\right) \mathrm{Re}\, S(E^* + i0^+). \tag{15}$$

As can be seen by comparing (11) and (14b), this function is periodic in E^*.

By combining (15) and (12) we can in principle calculate the exact density of states, in spite of the large value of δt. In practice however the sum in (13) must be cut off after a finite number, N, of steps. Then (14b) becomes

$$\mathrm{Re}\, S(Z) = 1 + \left\{ \frac{\sin[\frac{1}{2}(Z - E^*)\delta t(2N+1)]}{\sin[\frac{1}{2}(Z - E^*)\delta t]} \right\}. \tag{16}$$

When $\gamma = \mathrm{Im}\, Z$ is small the first zero of the numerator of (16) gives us a *generous* estimate of the width of the peak:

$$\delta E^* = \frac{4\pi}{2N+1} \approx \frac{2\pi}{T} \tag{17}$$

where T is the total integration time. Alternatively we can choose a finite value for γ so that $\delta E = \gamma$ is the required error in E. In general γ may be chosen to be a function of E which takes the effects of (11) into account. In practice it is unnecessary to integrate beyond

$$N\,\delta t = T \approx \frac{2\pi}{\gamma}. \tag{18}$$

(d) Summary

To end this section let us summarise the numerical procedure:

i) Choose $\psi(0)$ and calculate $\psi(\delta t)$ using some higher order, e.g. Runge-Kutta, scheme;
ii) Integrate using (6);
iii) Add the new contributions to the Laplace transforms (13);
iv) repeat (ii) and (iii) N times;
v) correct the calculated density of states using (12).

4. Broadening Functions

Experimental Spectra are generally broadened by convolution with some function related to (e.g.) particle lifetimes, properties of aparatus, etc. We have already discussed one form of broadening, namely the Lorentzian/Cauchy broadening due to the inclusion of the parameter γ. In general, however, we may include any broadening function by modifying (4a) to read

$$G(E) = \int_0^\infty \exp(iE\,t) g(t) b(t)\, dt \tag{19}$$

broadening function $B(E)$, and $B(E)$ is real (see Appendix A). In practice because of the large δt the actual broadening function will also be periodic in the same way as the density of states itself. This produces no effect which is not already present before the broadening and can therefore be ignored. In particular we have

$$b(t) = \exp(-|\gamma t|) \tag{20a}$$

$$b(t) = \exp(-\tfrac{1}{2}\sigma^2 t^2) \tag{20b}$$

for Lorentzian and Gaussian broadening respectively.

More complicated forms are also possible, including broadening functions $b(t)$ which depend on Z.

5. Comparison with the Recursion Method

Having established the general properties of the equation of motion method when applied to the calculation of densities of states, let us now make a detailed comparison with the recursion method which is the main subject of the present volume.

(a) Information Content

Both methods generate a set of functions ϕ_n which are linear combinations of the set $\mathcal{H}^n\phi_0$. In the case of the recursion method the set is orthogonal whereas those generated by the present method are not. They both span the same Hilbert space, however, and therefore carry in principle the same information.

We have already shown (18) that after N steps the resolution of the method is given by $1/N th$ of the total bandwidth. At the same stage in the recursion method the continued fraction gives N delta functions which may be broadened by finite γ or by the termination procedure. While the termination may make the spectrum look more appropriate for an infinite system it should not introduce any new information. Thus the genuine information the two methods give after N steps is identical. They differ only in the details of the smoothing.

(b) Termination and Smoothing

Much has been written on the subject of termination procedures for the recursion method [7,8], but very little on the equivalent problem for the equation of motion method, namely the long-time behaviour of $g(t)$. Actually the latter statement is untrue: a famous paper by P.W.Anderson [9] from 1958 discusses this very subject and gave rise to the concept of localisation in a random potential which has been the subject of so much interest in recent years. Indeed, almost the whole of the literature on localisation might be considered to be discussing $g(t)$ for large t.

The essential result of Anderson for disordered systems remains valid in principle although modified in detail. Localised states give a $g(t)$ which approaches a constant value asymptotically whereas extended states can be considered to be diffusive, so that

$$g(t \to \infty) \sim t^{-d/2} \tag{21}$$

where d is the dimensionality of the system.

The equivalent result for free electrons or for an ordered lattice is

$$g(t \to \infty) \sim t^{-d} \tag{22}$$

as derived in any textbook of quantum mechanics.

We have already discussed the convolution of the calculated density of states with a broadening function in the case of the equation of motion method. For the recursion method non-Lorentzian, e.g. Gaussian, broadening is much more difficult and has not to my knowledge been attempted.

(c) Timing

No comparison of computational methods is complete without a discussion of relative timing. Since we have already established that the steps in the two methods are roughly equivalent, we must now consider the time per step.

In the analysis of section (3) we have made use of the fact that the *leap-frog* method when applied to oscillatory equations conserves the normalisation of the wave functions. The Lanczos algorithm by contrast requires the renormalisation of the wave function at each stage in order to calculate the elements of the tridiagonal matrix. Indeed sometimes (e.g. on a highly parallel processor) the necessary scalar products may dominate the time required for each step. The calculation of $g(t)$ requires only one scalar product per step whereas the Lanczos algorithm requires two scalar products and the renormalisation of the wave functions.

We therefore conclude that the equation of motion method may just have the edge on timing.

6. Conclusion

The discussion of the equation of motion method presented here depends crucially on the intrinsic stability of the *leap-frog* algorithm for oscillatory equations. This allows us to choose a relatively large time step δt and then to compensate for the *systematic* error thus introduced. This aspect, which was not considered in previous comparisons between the equation of motion and recursion methods [4], enables us to achieve the same information per step in both methods. Since the steps are simpler in the equation of motion method this method is preferable for the calculation of densities of states.

The physical meaning of the time-development of a wave function is much clearer than that of the functions generated by the Lanczos algorithm. It is therfore more likely that other physical properties may be derived from an analysis of this behaviour.

I hope that the results presented here will stimulate further research on the equation of motion method and thus lead to more efficient calculations on large or disordered systems.

Acknowledgements

I should like to thank the Aspen Center for Physics where much of this work was done and the British Petroleum Venture Research Unit for financial support.

Appendix A

We require the imaginary part of $A(\omega)$ where

$$A(\omega) = \int_0^\infty \exp(i\omega t) a(t)\, dt. \tag{A1}$$

Then we may write

$$\mathrm{Im}\,A(\omega) = \tfrac{1}{2}\Big[\int_0^{\infty} \exp(i\omega t)a(t)\,dt - \int_0^{\infty} \exp(-i\omega t)a^*(t)\,dt\Big] \tag{A2a}$$

$$= \tfrac{1}{2}\int_{-\infty}^{+\infty} \exp(i\omega t)a(t)\,dt \tag{A2b}$$

iff $a(t) = a^*(-t)$. This condition is always true for $g(t)$ [see (3)].

Using the *Fourier* convolution theorem

$$\int_{-\infty}^{+\infty} A(\omega-\omega')B(\omega')\,d\omega' = \int_{-\infty}^{+\infty} a(t)b(t)\exp(i\omega t)\,dt \tag{A3}$$

and (A2b) we can write

$$\mathrm{Im}\int_0^{\infty} A(\omega-\omega')B(\omega')\,d\omega' = \mathrm{Im}\int_0^{\infty} a(t)b(t)\exp(i\omega t)\,dt \tag{A4}$$

as long as $a(t)b(t) = a^*(-t)b^*(-t)$.

References

[1] R.Alben, M.Blume, H.Krakauer, L.Schwartz: Phys.Rev.B**12**, 4090 (1975)
[2] D.Weaire, A.R.Williams: J.Phys.C**10**, 1239 (1977)
[3] B.Kramer, A.MacKinnon, D.Weaire: Phys.Rev.B**23**, 6357 (1981)
[4] D.Weaire, E.P.O'Reilly: J.Phys.C, (to be published)
[5] Other Authors: This Volume.
[6] D.E.Potter: "Computational Physics" (Wiley,1972) pp37–38.
[7] R.Haydock: Solid State Physics**35**, 215 (1980)
[8] C.M.M.Nex: J.Phys.A**11**, 653 (1978)
[9] P.W.Anderson: Phys.Rev.**109**, 1492 (1958)

Use of Cyclic Matrices to Obtain Analytic Expressions for Crystals

Philippe Audit

Laboratoire PMTM, Université Paris-Nord, F-93430 Villetaneuse, France

1. Introduction

Whereas the recursion method [1,2] allows to calculate the electronic structure of systems without any symmetry, a profitable use of cyclic matrix functions (CMF) requires the presence of some translational symmetry. Actually, the latter can be often more fully exploited using CMF rather than Bloch waves. The distinct advantages offered by CMF include in particular: first the possibility to deal directly with the physical properties expressible in terms of some element or the trace of a CMF, so bypassing the band structure calculation in situations where it is irrelevant. Second, it is applicable to functions of several cyclic matrices, not merely a single one (the Hamiltonian for instance), thereby offering an analytic solution for the difficulty associated with the nonorthogonality of the localized orbitals. Third, the assumption of an infinite lattice, an integral step in band structure calculations, appears to be superfluous here, as CMF can be defined and analytically expressed for lattices of finite size. Fourth, CMF exhibit a good flexibility to be perturbed by the presence of defects, in a way that is well referenced to bulk properties.

In the present survey of the CMF approach, I have laid stress on the ability of the method to yield analytic solutions when dealing with simple non self-consistent models. Apart from a better physical understanding of the electronic properties which can be gained from such results, analytical formulas provide an effective method of checking the accuracy of the various numerical methods available in more realistic situations. For example, the proposed analytic treatment of clusters' properties certainly offers interest by itself, but it provides also a useful test to calibrate the size of clusters required to model the bulk material in the recursion method [2].

The next section, where the main CMF properties , are summed up is self-contained without references to crystals, in order not to preclude other potential applications of the theory. The analytic expressions proposed for the CMF elements are in the form of Fourier series or Fourier integrals, according as the size of the lattice in all space directions is small or large. Mixed series-integral formulations are also available, which are useful to deal with lattices having different extensions, depending on the direction. Thus, the variations of the CMF elements induced by the passage from finite to infinite lattice size are exactly formulated. Moreover, the formalism enlightens the very simple relationship which connects the properties of finite and infinite lattices. In section III is proposed a cyclic matrix formulation of the LCAO method, which provides expressions in closed form for many electronic properties of the crystal and deals successfully with both difficulties respectively associated with a finite size lattice and a substantial amount of overlap between orbitals. Thereby is obtained a general solution for the static quantum size effects [4] in the LCAO approximation, as well as a rigorous estimation of how quickly the various electronic properties of the cluster converge to their bulk counterpart with increasing cluster size [5]. Section IV deals with situations which slightly depart from the perfect infinite lattice, namely tight-binding models of a vacancy in an atomic chain and of a surface in a semi-infinite three-dimensional crystal. In both cases, illuminating analytic results are obtained for the defect-induced charge transfer, for the rubbing out of Van Hove singularities in the density of states, and for other main electronic properties, which cast some light on the onset of disorder in a crystal.

2. Properties of Cyclic Matrix Functions

A cyclic matrix is the name for a square matrix whose each line (resp. column) can be generated from the preceding one by cyclic permutation of its elements. Consequently, the elements of any cyclic NxN cyclic matrix can be labelled by using a single indice ℓ, $\ell=0,1,\dots,N-1$, which refers to its column for an element belonging to the first line. Of particular interest are the topological NxN matrices [6,3] of order q, defined by

$$[m_q]_\ell = \begin{cases} 1 \text{ , if } q-\ell = 0,N,2N,\dots \\ 0 \text{ , otherwise} \end{cases} \tag{1}$$

m_{-q} is the transpose and inverse of m_q; they are commutative and satisfy the following relations

$$m_0 = m_N = I_N \text{ ,} \tag{2}$$

where I_N is the NxN unit matrix,

$$m_\alpha m_\beta = m_\beta m_\alpha = m_{\alpha+\beta} \text{ ,} \tag{3}$$

$$(m_\alpha)^p = m_{\alpha p} \text{ .} \tag{4}$$

Those matrices constitute a natural basis set for the representation of the whole class of cyclic NxN matrices.

Now, let us consider a simple cubic lattice of $N_1 x N_2 x N_3$ sites, with periodic boundary conditions; we are interested in generalized cyclic matrices in the form

$$A = \sum_{\vec{p}\in P} m_{p_1} \otimes m_{p_2} \otimes m_{p_3} \otimes a_{\vec{p}} \text{ ,} \tag{5}$$

$$= \sum_{\vec{p}\in P} (m_1 \otimes I \otimes I)^{p_1} \, (I \otimes m_1 \otimes I)^{p_2} \, (I \otimes I \otimes m_1)^{p_3} \otimes a_{\vec{p}} \tag{6}$$

where $a_{\vec{p}}$ is a matrix describing some interaction existing between the sites $\vec{0}$ and $\vec{p} = (p_1,p_2,p_3)$. The summation extends over a set P of sites surrounding the origin (or any site); for a limited range of interaction, we have $P = \{\vec{p} \,|\, p_i \leq P_i\}$, otherwise P coincides with the SC lattice; by turning off some interactions, i.e. setting some $a_{\vec{p}}$'s equal to zero, one obtains a FCC or a BCC lattice. Then, let us turn to functions of the generalized cyclic matrix A. The elements of f(A) are matrices having the same dimension than $a_{\vec{p}}$, which can be conveniently labelled by using the real-space vector $\vec{\ell}=(\ell_1,\ell_2,\ell_3)$.

Several analytical expressions for those matrix elements have been documented elsewhere [3]. They are designed to deal adequately with lattices of different shapes. Considering first the case of a lattice infinite in the three directions of space, and denoting by A^∞ the matrix (5) defined on such a lattice, we may build from (6) the following matrix

$$a(\vec{\theta}) = \sum_{\vec{p}\varepsilon P} a_{\vec{p}} \, e^{i\vec{p}.\vec{\theta}} \quad , \tag{7}$$

where $\vec{\theta}$ is a reciprocal vector with three continuous components $(\theta_1,\theta_2,\theta_3)$, $-\pi\leqslant\theta_i\leqslant\pi$. Then the matrix elements are obtained in the form of the Fourier integral

$$[f(A^{\infty})]_{\vec{\ell}} = (8\pi^3)^{-1} \int_{-\pi}^{\pi} d\vec{\theta} \; f(a(\vec{\theta})) \; e^{-i\vec{\ell}\cdot\vec{\theta}} \quad . \tag{8}$$

This result turns out to be also useful for a CMF f(A) defined on a parallelepiped consisting of $N_1N_2N_3$ sites respectively in the three directions of space, as we have

$$[f(A)]_{\vec{\ell}} = \sum_{\vec{j}} [f(A^{\infty})]_{\vec{\ell}+\vec{j}} \quad , \tag{9}$$

where $\vec{j} = (j_1N_1,j_2N_2,j_3N_3)$, j_i integer, defines a site in the superlattice whose unit cell is the parallelepiped of sides N_1,N_2,N_3. Practically the summation over $\vec{j}$ is limited to a few terms for large N_i values. More precisely, f(A) being defined by a power series, for a term of order q must be satisfied the inequality: $j_iN_i \leqslant qP_i-\ell_i$ that determines the restricted range of summation over $\vec{j}$, according to the series convergence rate.

But the lattice could also extend infinitely in only one or two directions; that is a situation intermediate between (8) and (9). For example, a thick slab with N layers allows the following expression

$$[f(A_s)]_{\vec{\ell}} = (8\pi^3)^{-1} \sum_{j=0}^{\infty} \iiint_{-\pi}^{\pi} d\theta_1 d\theta_2 d\theta_3 f(a(\vec{\theta}))e^{-i(\ell_1\theta_1+\ell_2\theta_2+(\ell_3+jN)\Theta_3)} \quad ,$$

$$= \sum_{j=0}^{\infty} [f(A^{\infty})]_{\ell_1,\ell_2,\ell_3+jN} \tag{10}$$

it gives the result in terms of only a few (cf.the above discussion) infinite matrix elements, which are available in analytic form, from (8).

Alternatively, defining a reciprocal vector with discrete components

$$\vec{\tau} = 2\pi \left(\frac{\tau_1}{N_1} , \frac{\tau_2}{N_2} , \frac{\tau_3}{N_3} \right) \quad , \quad 0 \leqslant \tau_i \leqslant N_i-1 \tag{11}$$

the CMF elements are expressible as the Fourier series

$$[f(A)]_{\vec{\ell}} = (N_1N_2N_3)^{-1} \sum_{\vec{\tau}} f(a(\vec{\tau}) \; e^{-i\vec{\ell}.\vec{\tau}} \quad , \tag{12}$$

which is most convenient to deal with lattices of small size.

Mixed components for the reciprocal vector can be considered as well ; for example in the case of a thin slab with N layers we have

$$[f(A_s)]_{\vec{\ell}} = (4\pi^2N)^{-1} \sum_{\tau=0}^{N-1} e^{-\frac{2\pi i\tau\ell_3}{N}} \iint_{-\pi}^{\pi} d\theta_1 d\theta_2 f(a(\theta_1,\theta_2,\tfrac{2\pi\tau}{N}))e^{-i(\ell_1\theta_1+\ell_2\theta_2)} \tag{13}$$

where the reciprocal vector is $\vec{t} = (\theta_1 , \theta_2 , \frac{2\pi\tau}{N})$; the expression (13) is a sum of N two-dimensional CMF elements which happens to be more adequate than (10) when N is small.

The trace of the CMF is another very useful quantity that is straightforward to derive upon setting $\vec{\tau}=\vec{0}$ in the former expressions. We have in the limit $N_1,N_2,N_3 \to \infty$

$$(N_1N_2N_3)^{-1}\mathrm{Tr}\{f(A^\infty)\}= (8\pi^3)^{-1}\int_{-\pi}^{\pi} d\vec{\theta}\ \mathrm{Tr}\{f(a(\vec{\theta}))\} \quad , \tag{14}$$

$$\mathrm{Tr}\ f(A) = N_1N_2N_3 \sum_{\vec{j}} \mathrm{Tr}\ [f(A^\infty)]_{\vec{j}} \quad , \tag{15}$$

and also in the limit $N_1,N_2 \to \infty$

$$(N_1N_2)^{-1}\mathrm{Tr}\ f(A_s) = \sum_{j=0}^{\infty} \mathrm{Tr}\ f(A^\infty)_{0,0,jN} \tag{16}$$

$$= (4\pi^2)^{-1}\sum_{\tau=0}^{N-1} \iint_{-\pi}^{\pi} d\theta_1 d\theta_2 f(a(\theta_1,\theta_2,\tfrac{2\pi\tau}{N})) \quad . \tag{17}$$

For <u>infinite isotropic lattices</u>, the cyclic matrices used to describe the isotropic interactions taking place in the lattice should preferably be written in terms of matrices in the form

$M_p = m_p + m_{-p}$,

which is the symmetric topological cyclic matrix of order p, so as to take advantage of a useful property [3,6]

$$M_p = C_p(M_1) \quad , \tag{18}$$

where $C_p(x)$ is the pth order Chebysheff polynomial of the second kind [7] , obeying the recurrence formula

$$C_{p+1}(x) = xC_p(x) - C_{p-1}(x) \ , \qquad p \geqslant 1 \tag{19}$$

$$C_0(x) = 2 \qquad , \qquad C_1(x) = x \ .$$

Taking advantage of the orthogonal properties of Chebysheff polynomials, we can express any CMF defined on a one-dimensional infinite lattice in the form

$$f(A) = \sum_{q=0}^{\infty} \langle f(x),C_q(x)\rangle\ C_q(A) \quad , \tag{20}$$

and specially

$$f(M_1) = \sum_{q=0}^{\infty} \langle f(x),C_q(x)\rangle M_q \quad , \tag{21}$$

where the inner product is

$$\langle f(x),C_q(x)\rangle = (2\pi(1+\delta_q))^{-1} \int_{-1}^{1} (1-x^2)^{-1/2} f(2x)\ C_q(2x)\ dx \tag{22}$$

$$= (\pi(1+\delta_q))^{-1} \int_0^{\pi} f(2\cos\theta)\ \cos q\theta\ d\theta \ . \tag{23}$$

Generalization to three dimensions is obtained through the following formulas:

$$f(A_1\otimes A_2\otimes A_3)= \sum_{q_1,q_2,q_3=0}^{\infty} \langle f(x_1,x_2,x_3),C_{q_1}(x_1)C_{q_2}(x_2)C_{q_3}(x_3)\rangle\ C_{q_1}(A_1)\otimes C_{q_2}(A_2)\otimes C_{q_3}(A_3) \quad , \tag{24}$$

with the inner product

$$\langle f(x_1,x_2,x_3),C_{q_1}(x_1)C_{q_2}(x_2)C_{q_3}(x_3)\rangle= \left(\pi^3(1+\delta_{q_1})(1+\delta_{q_2})(1+\delta_{q_3})\right)^{-1} \iiint_0^{\pi} d\theta_1 d\theta_2 d\theta_3 f(2\cos\theta_1,2\cos\theta_2,2\cos\theta_3)\cos q_1\theta_1 \cos q_2\theta_2 \cos q_3\theta_3 \ . \tag{25}$$

3. The LCAO Method Revisited

The traditional approach of the LCAO method in solids involves a prerequisite calculation of the spectrum; from which, neglecting overlap, the density of states (DOS) for an infinite perfect crystal non-degenerate s band has been obtained analytically [8] ; whereas approximate formulas are available for some d-band models, see ref. [9] for an example. But in the presence of strong overlap, the use of a set of Bloch functions precludes full analytic treatment and causes trouble in numerical calculations [10]. For finite crystals the electronic structure is acknowledged to be much more complicated, and only a few interesting results in closed form have been obtained for some simple s band models of clusters [5,11].

In the models that I will discuss now, the atoms retain their bulk positions and satisfy periodic boundary conditions, but may exhibit most distant electronic hopping and overlap. The analytic treatment displays fully the effects of orbitals' non-orthogonality and crystal size on the LCAO formalism. Note that a similar analysis could describe phonons and magnons properties.

Let us consider a column matrix Φ , whose entries are the atomic orbitals (or other adequate localized functions) of type α centred on each lattice site $\vec{\ell}$ of the lattice; as they are normalized but generally not mutually orthogonal on different sites, an overlap matrix I+S whose elements are the overlap integrals $\langle\alpha\vec{\ell}|\beta\vec{n}\rangle$ can be defined; so that the Schrödinger equation can be approximated by [12]

$$(H-E(I+S))\Phi = 0 \quad . \tag{26}$$

This equation allows two interesting reformulations as follows

$$\left((I+S)^{-1}H - E\right)\Phi = 0 \quad , \tag{27}$$

$$\left((I+S)^{-1/2}H(I+S)^{-1/2} - E\right)(I+S)^{1/2}\Phi = 0 \quad . \tag{28}$$

These lead one to define a couple of effective Hamiltonians in the form

$$H' = (I+S)^{-1} H \quad , \tag{29}$$

$$H'' = (I+S)^{-1/2} H (I+S)^{-1/2} \quad . \tag{30}$$

The Hamiltonian and the overlap matrix have the same generalized cyclic matrix structure

$$H = \sum_{\vec{p}\in P} m_{p_1} \otimes m_{p_2} \otimes m_{p_3} \otimes h_{\vec{p}} \quad , \tag{31}$$

$$S = \sum_{\vec{p}\in P}' m_{p_1} \otimes m_{p_2} \otimes m_{p_3} \otimes s_{\vec{p}} \quad , \tag{32}$$

where the prime on the summation denotes restriction to $\vec{p} \neq \vec{0}$. The entries of $h_{\vec{p}}$ and $s_{\vec{p}}$ are respectively the hopping and the overlap integrals between orbitals belonging to atoms in sites $\vec{0}$ and $\vec{p}$. From (31) and (32) follow the matrices

$$h(\vec{t}) = \sum_{\vec{p}\in P} e^{i\vec{p}.\vec{t}}\, h_{\vec{p}} \quad , \tag{33}$$

$$s(\vec{t}) = \sum_{\vec{p}\in P} e^{i\vec{p}.\vec{t}}\, s_{\vec{p}} \quad , \tag{34}$$

that are needed to build matrices

$$h'(\vec{t}) = (I+s(\vec{t}))^{-1} h(\vec{t}) \quad , \tag{35}$$

$$h''(\vec{t}) = (I+s)^{-1/2} h\, (I+s)^{-1/2} \quad , \tag{36}$$

$$= \sum_{\vec{p}} h_{\vec{p}} e^{i\vec{p}.\vec{t}} - \frac{1}{2}\sum_{\vec{p}}\sum_{\vec{p}'}{}' (s_{\vec{p}'} h_{\vec{p}} + h_{\vec{p}} s_{\vec{p}'}) e^{i(\vec{p}+\vec{p}').\vec{t}} + \ldots \quad , \tag{37}$$

to be used in calculating functions of the effective Hamiltonians (29,30); and also the matrix

$$(I+s)^{1/2} = I + \frac{1}{2}s - \frac{1}{8}s^2 + \frac{1}{16}s^3 - \ldots \quad , \tag{38}$$

from which the new basis set $(I+S)^{1/2}\Phi$ in (28) can be obtained.

Now useful functions of the effective Hamiltonian are attainable through the different expressions of the CMF given in section 2, for various crystal shapes. For example, the exact Green's function for the effective Hamiltonian H', which is approximate due to the finiteness of the basis set [13], is readily derived with elements in the form

$$[G(E)]_{\vec{\ell}} = (2\pi)^{-3} \int_{-\pi}^{\pi} (E - h'(\vec{\theta}))^{-1} e^{-i\vec{\ell}.\vec{\theta}} \, d\vec{\theta} \quad . \tag{39}$$

Subsequently, the local DOS [14] at an orbital is derived

$$n_\alpha(E) = -\pi^{-1} \operatorname{Im}[G(E+i0)]_{\vec{0},\alpha\alpha} \quad ,$$

$$= (2\pi)^{-3} \sum_n \int_{-\pi}^{\pi} \frac{[\operatorname{adj}(E_n - h')]_{\alpha\alpha}}{\operatorname{Tr}\{\operatorname{adj}(E_n - h')\}} \delta(E - E_n) \, d\vec{\theta} \quad , \tag{40}$$

where adj(A) denotes the adjoint of the matrix A, and E_n is an eigenvalue of the matrix h'. The total DOS follows in the form

$$n(E) = (2\pi)^{-3} \sum_n \int_{-\pi}^{\pi} \delta(E - E_n(\vec{\theta})) \, d\vec{\theta} \quad . \tag{41}$$

The one-electron energy per site, which is formally defined by integration over the occupied part of the spectrum as

$$U = N^{-1} \sum_n E_n f(E_n) = \int_{-\infty}^{\infty} E \, n(E) \, f(E) \, dE \quad , \tag{42}$$

where f is the Fermi occupation factor: $f(E) = (1 + \exp\beta(E-\mu))^{-1}$, is also obtained as

$$U = N^{-1} \operatorname{Tr}\{H' f(H')\}$$

$$= (2\pi)^{-3} \int_{-\pi}^{\pi} \operatorname{Tr}\{h'(I + \exp\beta(h'-\mu))^{-1}\} d\vec{\theta} \quad . \tag{43}$$

In a similar way, other thermodynamical functions can be attained, such as free energy or electronic specific heat at low temperature. This is given as:

$$C_v/k = \beta^2 (2\pi)^{-3} \int_{-\pi}^{\pi} \operatorname{Tr}\{(h'-\mu)(I + \exp\beta(h'-\mu))^{-2} \exp\beta(h'-\mu)\} \, d\vec{\theta} \quad . \tag{44}$$

Notice that the above results in terms of H' are also easily expressible in terms of H" by using (37) instead of (35).

For illustration, I consider now the simple example of a CC crystal s-band with nearest neighbours overlap and hopping; the matrices which reduce to scalars in that case read

$$h(\vec{\theta}) = E_0 - 8E_1 \cos\theta_1 \cos\theta_2 \cos\theta_3 \quad ,$$

$$s(\vec{\theta}) = 8\sigma \cos\theta_1 \cos\theta_2 \cos\theta_3 \quad ,$$

$$h'(\vec{\theta}) = \frac{E_0 - 8E_1 \cos\theta_1 \cos\theta_2 \cos\theta_3}{1 + 8\sigma \cos\theta_1 \cos\theta_2 \cos\theta_3} \quad ;$$

using (41) the DOS, for example, is easily derived in the form

$$n(E) = \pi^{-2} \frac{1+\sigma E_0/E_1}{1+\sigma E/E_1} \iint_0^{\pi} \{\cos^2\theta_1\cos^2\theta_2 - [\frac{E-E_0}{8(E_1+\sigma E)}]^2\}^{-1/2} d\theta_1 d\theta_2 \quad ,$$

which clearly shows the influence of the overlap : the band edges are shifted as

$$\frac{E_0-8E_1}{1+8\sigma} \leqslant E \leqslant \frac{E_0+8E_1}{1-8\sigma} \quad ;$$

and the band width becomes $\dfrac{16(E_1+\sigma E_0)}{1-64\,\sigma^2}$.

Therefore, based on (8), a general formulation has been obtained for the main electronic properties of an infinite cubic crystal, in the LCAO approximation, which takes fully into account the orbital overlap.

To extend the above results to crystals of finite size, rather than (8) which is valid for infinite crystals, the adequate general expression (9,10, 12, or 13) for the CMF elements should be chosen according to the shape of the crystal under consideration. In fact, from section 2, we have two main formulation at our disposal. In the first one, it follows from equations (9),(10),(15) and (16), that the finite crystal properties can be expressed in terms of the infinite crystal matrix elements, which is most convenient for large N_i, when just a few terms are sufficient to describe the perturbation caused by the limitation of size. The transformation

$$\ell_i \to \ell_i + j_i N_i \qquad \int_{-\pi}^{\pi} \to \sum_{j_i=0}^{\infty} \int_{-\pi}^{\pi} \tag{45}$$

allows to extend to the finite crystal, the validity of the infinite crystal expressions discussed above in the present section.

A second possible formulation results from the equations (12),(13),(17). Here the matrix elements exhibit clearly oscillations as a function of the crystal thickness, that are similar to those previously studied for a jellium model under the name of Quantum Size Effect [4]. The following transformation can be used to obtain the properties of the finite crystal with small N_i from the infinite crystal properties:

$$\theta_i \to \frac{2\pi\tau_i}{N_i} \quad , \qquad (2\pi)^{-1}\int_{-\pi}^{\pi} d\theta_i \to N_i^{-1} \sum_{\tau_i=0}^{N_i-1} \tag{46}$$

For illustration, let us consider the DOS for the s-band of a thin slab with nearest neighbour interactions. For an orthogonal basis and a proper choice of the energy's origin, the equation (33) takes the form

$$h(\vec{t}) = 2E_1(\cos\theta_1 + \cos\theta_2 + \cos\frac{2\pi\tau}{N}) \quad . \tag{47}$$

Then from (41) and (46) the DOS per site of the slab is found to be

$$n_S(E) = (4\pi^2 N)^{-1} \sum_{\tau=0}^{N-1} \iint_{-\pi}^{\pi} \delta\{E-2E_1(\cos\frac{2\pi\tau}{N} - 2E_1(\cos\theta_1+\cos\theta_2)\}\, d\theta_1 d\theta_2 \, . \tag{48}$$

Moreover, the DOS for an infinite square lattice being

$$n_2(E) = (4\pi^2)^{-1} \iint_{-\pi}^{\pi} \delta\{E-2E_1(\cos\theta_1+\cos\theta_2)\}\, d\theta_1 d\theta_2 \quad , \tag{49}$$

equation (48) can be rewritten as

$$n_S(E) = N^{-1} \sum_{\tau=0}^{N-1} n_2(E-2E_1\cos\frac{2\pi\tau}{N}) ,$$

where the DOS of the slab appears to be equal to a sum of "translated" square lattice DOS.

4. Matrix Functions Formalism for Imperfect Crystals

A direct and natural description of crystals with defects is offered by the representation of the tight-binding Hamiltonian in terms of some combination of generalized cyclic matrices and sparse diagonal matrices. Then the consideration of various functions of the Hamiltonian,which diverge more or less from a typical CMF according to the degree of disorder, can yield some analytic closed expressions for the imperfect crystal electronic properties, without appealing to the Bloch's theorem. Two simple examples will be considered,in order to show the feasibility of the method.

The diagonal NxN matrices D_p used to describe the defects are defined in the following manner

$$[D_p]_{ij} = \delta_{ij}\,\delta_{ip} \quad , \; p\geqslant 0 \tag{50}$$

$$[D_{-p}]_{ij} = [D_{N-p}]_{ij} = \delta_{ij}\,\delta_{i,N-p} \quad , \quad p\geqslant 0 \tag{51}$$

they clearly satisfy the following identities

$$D_p^{\,2} = D_p \quad , \quad D_p\,D_q = 0 \quad \text{if} \quad p\neq q \;. \tag{52}$$

Furthermore, the matrices m_p and D_p do not commute, but rather obey the rule

$$D_p m_q = m_q D_{p+q} \quad \text{or} \quad m_q D_p = D_{p-q} m_q \;. \tag{53}$$

The Hamiltonian for a tight-binding model of an <u>infinite-chain with a vacancy</u> located at the origin, with periodic boundary conditions and nearest neighbours interactions only, can be written as

$$H_v = \varepsilon(I-D_0)+\beta((I-D_0-D_{-1})m_1+m_{-1}(I-D_0-D_{-1})) \tag{54}$$

where ε is the atomic energy at the occupied sites and β the transfer energy. Changing the zero and scale of energy, it is convenient to use the reduced Hamiltonian

$$H = \beta^{-1}(H_v-\varepsilon I) = \alpha D_0+(I-D_0-D_{-1})m_1+m_{-1}(I-D_0-D_{-1}) \quad , \quad \alpha=-\varepsilon/\beta \;. \tag{55}$$

The matrix H is not much different from the matrix M_1 which obeys the simple relation (21); this statement leads one naturally to use the expansion (20) to determine f(H). In the present case, the matrix Chebysheff polynomials $C_q(H)$ obey recurrence relations which allow an exact summation of (20), whose first order terms are in the form

$$f(H) = 2\langle f(x),C_0(x)\rangle(I-D_0)+f(\alpha)D_0-\sum_{\ell=1}^{(N-1)/2}\langle f(x),C_{2\ell}(x)\rangle(D_\ell+D_{-\ell})$$
$$+\langle f(x),C_1(x)\rangle m_1-\sum_{\ell=0}^{N/2-1}\langle f(x),C_{2\ell+1}(x)\rangle(D_\ell+D_{-\ell-1})m_1- \dots \;. \tag{56}$$

This general relation, being applied to the determination of the matrix resolvent

$$G(z) = (zI-H)^{-1} ,$$

yields the following expressions for its elements

$$G_{00}(z) = (z-\alpha)^{-1} ,$$

$$G_{\ell\ell}(z)=G_{N-\ell,N-\ell}(z)=<(z-x)^{-1},2C_0(x)-C_{2\ell}(x)>= \frac{2}{\pi}\int_0^{\pi} \frac{\sin^2\ell\theta \, d\theta}{z-2\cos\theta} \qquad (\ell\neq 0)$$

$$G_{\ell,\ell+1}=<(z-x)^{-1},C_1(x)-C_{2\ell+1}(x)>=\pi^{-1}\int_0^{\pi} \frac{\cos\theta-\cos(2\ell+1)d\theta}{z-2\cos\theta} ,$$

by taking into account the inner product definition (23).

The local DOS follows from those last results, in the form

$$n_0(E) = \delta(E-\alpha)$$

$$n_\ell(E)=n_{N-\ell}(E)=<\delta(E-x),2C_0(x)-C_{2\ell}(x)> = \frac{1-C_{2\ell}(E)/2}{(1-E^2/4)^{1/2}} ;$$

and the total density of states has the simple form

$$n(E) = N\pi^{-1}(1-E^2/4)^{-1/2}+\delta(E-\alpha)-\frac{1}{2}\left[\delta(E-2)+\delta(E+2)\right] . \qquad (57)$$

This analytical expression of the spectrum involves three different contributions: in the first term, the DOS of the perfect chain is retrieved; the second term represents a localized bound state at zero energy, which is known to be caused by a diagonal perturbation [15,16]; whereas the two negative δ functions in the third term, that tend to rub out the Van Hove singularities at the band edges, result from the breaking of the neighbouring bond pair of the vacancy, so illustrating a more general result [17].

The diagonal terms of the matrix (56) can be added together to give the trace in the form

$$\mathrm{Tr}\, f(H) = 2N<f(x),C_0(x)>+f(\alpha)-\frac{1}{2}\{f(2)+f(-2)\} . \qquad (58)$$

Making use of this general expression, the spectrum calculation may be irrelevant; in particular, thermodynamical functions can be obtained directly. For example, the free energy with respect to the chemical potential μ is

$$F = -\beta \mathrm{Tr}\, \mathrm{Log}\{1+\exp\beta\, (\mu-H)\} ; \qquad (59)$$

and the variation in the free energy due to the vacancy can be written in the form

$$\Delta F=-\beta \mathrm{Log}\{\left[1+\exp \beta(\mu-\alpha)\right]\left[1+\exp\beta(\mu-2)\right]^{-1/2}\left[1+\exp \beta(\mu+2)\right]^{-1/2}\}. \qquad (60)$$

Now, my second example concerns the electronic structure of a semi-infinite crystal. By considering the non-selfconsistent tight-binding model of a non-degenerate band, with nearest-neighbour interactions in a SC crystal having broken bonds between two adjacent (100) planes, I refer to an old problem, which has been already explored in the past by the moment method [18] and the Green's function method [19]. In fact the CMF appears to compare favourably with previous approaches by exhibiting more ability to yield analytic results.

The Hamiltonian of the one-dimensional lattice with one broken bond between sites 0 and 1 is in the form

$$H_1 = (I-D_0)m_1+m_{-1}(I-D_0) , \qquad (61)$$

and the Hamiltonian of the SC crystal with broken bonds between planes (100) reads

$$H_S = I\otimes I\otimes H_1 + I\otimes M_1\otimes I + M_1\otimes I\otimes I \quad . \tag{62}$$

Following the same procedure as in the preceding example, and taking into account (18),(24) , a function of the Hamiltonian can be written as

$$f(H_S) = \sum_{q_1,q_2,q_3=0}^{N-1} \langle f(x_1+x_2+x_3), C_{q_1}(x_1)C_{q_2}(x_2)C_{q_3}(x_3)\rangle M_{q_3}\otimes M_{q_2}\otimes C_{q_1}(H_1) \; .$$

Making use of the rules (2,3,4) and (52,53), recurrence relations have been derived for matrices $C_q(H_1)$ which allow one to rewrite the latter equation in the form

$$f(H_S) = \sum_{q_2,q_3=0}^{N-1} \langle f(x_1+x_2+x_3), C_0(x_1)C_{q_2}(x_2)C_{q_3}(x_3)\rangle M_{q_3}\otimes M_{q_2}\otimes 2I$$

$$- \sum_{q_1=1}^{(N-1)/2} \sum_{q_2,q_3=0}^{N-1} \langle f(x_1+x_2+x_3), C_{2q_1}(x_1)C_{q_2}(x_2)C_{q_3}(x_3)\rangle M_{q_3}\otimes M_{q_2}\otimes (D_{q_1}+D_{-q_1+1})$$

$$+ \sum_{q_2,q_3=0}^{N-1} \langle f(x_1+x_2+x_3), C_1(x_1)C_{q_2}(x_2)C_{q_3}(x_3)\rangle M_{q_3}\otimes M_{q_2}\otimes (M_1-D_0m_1-m_{-1}D_0)$$

$$- \sum_{q_1=1}^{(N-1)/2} \sum_{q_2,q_3=0}^{N-1} \langle f(x_1+x_2+x_3), C_{2q_1+1}(x_1)C_{q_2}(x_2)C_{q_3}(x_3)\rangle M_{q_3}\otimes M_{q_2}\otimes \Big[(D_{q_1}+D_{-q_1})m_1$$

$$+m_{-1}(D_{q_1}+D_{-q_1})\Big] + \; .... \; , \tag{63}$$

that is here merely limited to the first off-diagonal elements for simplicity.

The elements of the Green's function are readily obtained from (63) in the form

$$G(\ell\ell'\ell'',\ell\ell'\ell'';z) = G(\ell 00,\ell 00;z)$$

$$= \frac{2}{\pi^3} \iiint_0^{\pi} \frac{\sin^2 \ell\theta_1 \; d\theta_1 d\theta_2 d\theta_3}{z-2\cos\theta_1-2\cos\theta_2-2\cos\theta_3} \qquad (\ell\neq 0) \; ; \tag{64}$$

and similarly for the off-diagonal elements as

$$G(\ell_1\ell_2\ell_3,\ell_1+1\;\ell_2'\ell_3';z) = \frac{1}{\pi^3}\iiint_0^{\pi} \frac{\left[\cos\theta_1-\cos(2\ell_1+1)\theta_1\right]\cos(\ell_2'-\ell_2)\theta_2\cos(\ell_3'-\ell_3)\theta_3}{z-2\cos\theta_1-2\cos\theta_2-2\cos\theta_3}$$

$$d\theta_1 d\theta_2 d\theta_3 \quad . \tag{65}$$

Denoting by $\nu_2(E)$ and $\nu_3(E)$ respectively the two-and three-dimensional perfect crystal density of states, the local density of states is found from (63) to be for the semi-infinite crystal in the form

$$n_{\ell 00}(E) = \nu_3(E)-4\langle\delta(E-x_1-x_2-x_3), C_{2\ell}(x_1)C_0(x_2)C_0(x_3)\rangle \quad ; \tag{66}$$

whereas the total DOS obeys the illuminating relation

$$n(E) - \nu_3(E) = N^{-1}\left[\nu_3(E)-\tfrac{1}{2}\{\nu_2(E+2)+\nu_2(E-2)\}\right] \tag{67}$$

where the Van Hove singularities of $\nu_3(E)$,arising at $E=\pm 2,\pm 6$, are doomed to be erased by the sharp band edges of $\nu_2(E\pm2)$ which coincide with them and have opposite sign.

The function $f(H_S)$ has, from (63), a trace in the form

$$\mathrm{Tr}\, f(H_S) = 8N^2(N+1)<f(x_1+x_2+x_3), C_0(x_1)C_0(x_2)C_0(x_3)>-2N^2<f(2+x_2+x_3) +f(-2+x_2+x_3), C_0(x_2)C_0(x_3)> , \quad (68)$$

which enables us to calculate the thermodynamic functions. The variation of the free energy due to the presence of the surface, for example, is written as

$$\Delta F = -\frac{N^2}{2\pi^2}\int\int_0^\pi \mathrm{Log}\{[1+\exp(\mu-2-2\cos\theta_1-2\cos\theta_2)][1+\exp(\mu+2-2\cos\theta_1-2\cos\theta_2)]\}d\theta_1 d\theta_2 .$$

In order that analytic results could be obtained easily, we have deliberately chosen simple models of defects. The ability of the formalism extends beyond those simple cases, as CMF show in fact more flexibility than Bloch's waves to be perturbed by the presence of defects.

References

1. R.Haydock,V.Heine,M.Kelly,J.Phys.C5,2845(1972)
2. R.Haydock in Solid State Physics, ed. by Ehrenreich,F.Seitz,D.Turnbull (Acad.Press, N.Y.1980), vol.35,p.215.
3. Ph.Audit,J.Physique 42,903(1981); and to be published
4. F.K.Schulte,Surf.Sci.55,427(1976)
5. R.P.Messmer,Phys.Rev.B15,1811(1977)
6. P.O.Löwdin,R.Pauncz,J.deHeer,J.Math.Phys.1,461(1960)
7. M.Abramowitz,J.Stegun,Handbook of Mathematical Functions (Dover 1970)
8. R.Jellitto,Phys.Chem.Solids,30,609(1969)
9. T.Wolfram,S.Ellialtioglu,Phys.Rev.B25,2697(1982)
10. T.Ahlenius,J.L.Calais,P.O.Löwdin,J.Phys.C6,1896(1973)
11. O.Bilek,L.Skala,L.Künne,Phys.Stat.Sol.b17,675(1983)
12. P.O.Löwdin,Adv.Phys.5,1(1956);Adv.Quant.Chem.5,185(1970)
13. A.R.Williams,P.J.Feibelman,N.D.Lang,Phys.Rev.B26(1982)
14. J.Friedel,Adv.Phys.3,446(1954)
15. E.W.Montroll,P.B.Potts,Phys.Rev.100,525(1955)
16. E.N.Economou,Green's functions in Quantum Physics(Springer-Verlag,Berlin 1983)
17. Ph.Audit,Phys.Rev.B30,(1 Oct.1984)
18. F.Cyrot-Lackmann,Surf.Sci.15,535(1968)
19. G.Allan,Ann.Phys.5,169(1970)

Part IV

Solid State Applications

Continued Fractions and Perturbation Theory: Application to Tight Binding Systems

P. Turchi+ and F. Ducastelle

Office National d'Etudes et de Recherches Aérospatiales (ONERA), B.P. 72
F-92322 Chatillon Cédex, France

1. Introduction

The electronic structure of transition metals and alloys has been widely investigated since the 70's through real space techniques and in particular by using the recursion method [1], well known for its numerical stability. Despite the success of this method that allows us to calculate band energies and to correctly reproduce general features and trends [2,3], its basic non linearity makes the results rather opaque. So the purpose of this paper is to present a simplification of the real space approach in order to recover the stability properties of transition metals and alloys in a more transparent way.

From a mathematical point of view, the Linearized Green function Method (LGM) we present hereafter is based on the fact that the density of states (DOS) and more particularly the band energy differences can be expanded as a sum of universal functions defined for a convenient reference medium, multiplied by fluctuations of continued fraction coefficients. These fluctuations can themselves be related to the differences between the moments of the DOS ; since these moments give an insight into the role of the d symmetry of atomic orbitals and of topological effects, we have a tool to make a "chemical" approach of the stability properties.

Furthermore, although the total energy of an ordered transition alloy cannot be expanded in terms of pair interactions, recent works [4] have shown that the configurational contribution can be written in this way. The LGM leads to a similar result and allows us to determine in an approximate way the most significant effective pair interactions. The general properties of these interactions are recovered and can be compared semi-quantitatively with those derived from the Generalized Perturbation Method [4-6] introduced previously.

As a matter of fact, the LGM can also be presented as an interesting method for studying amorphous versus crystalline metals and alloys through a careful analysis of the fluctuations of continued fraction coefficients.

Finally, from a practical point of view, the LGM can be successfully applied to the calculation of band energy, without loss of accuracy, by a judicious choice of the reference medium.

+Also with: Laboratoire de Dynamique du Réseau et Ultra-Sons
Université Pierre et Marie Curie, Tour 22,
4 Place Jussieu, 75005 Paris, FRANCE

The paper is organized as follows : in § 2 we outline the LGM and give its properties. In particular the convergence aspect is tackled, in order to clarify its applicability range. In § 3 we apply the LGM to the case of transition metals, and show very briefly the possible relations between phase-stability properties and topology, via a study of the moments of the DOS. The method is illustrated by the comparison between FCC and BCC structures. In § 4 we derive from a moment calculation and the LGM a general approximate expression for effective ordering pair interactions in transition alloys. Finally § 5 is devoted to a short conclusion.

2. The Linearized Green function Method : description and properties

2.1 Description of the LGM :

We assume the reader to be familiar with the continued fraction formalism applied to physical systems described by a hamiltonian H [7]. Then let us consider the matrix element of the resolvent defined at the first site of the semi-infinite linear chain generated by the recursion method and associated with the halmitonian H projected on the recursion basis $|n\}$:

$$\{1\,|G(z)|\,1\} = \{1|\,(z-H)^{-1}|\,1\} \equiv G_{11}(z)$$

where $G_{11}(z)$ is written in the form of a JACOBI continued fraction :

$$G_{11}(z) = \frac{1\,|}{|\,z-a_1} - \frac{b_1^2\,|}{|\,z-a_2}$$

The real coefficients (a_n, b_n^2) satisfy the well - known three term recurrence relation :

$$b_n\,|\,n+1\} = H|n\} - a_n|n\} - b_{n-1}|\,n-1\}$$

so that the representation of H in this basis can be illustrated by the following chain model

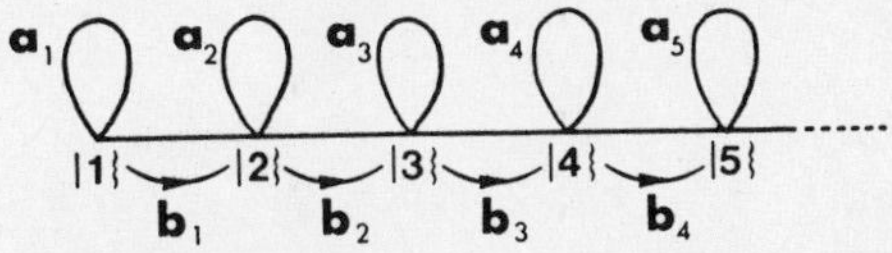

which will define our reference medium.

The associated DOS is as usual given by :

$$n(E) = \frac{-\mathrm{Im}}{\pi} \lim_{\eta \to 0} G_{11}(E+i\eta)$$

Without loss of generality we suppose that the initial hamiltonian H is, for example, a tight binding one written in the s band approximation.

We intend to evaluate the following difference :

$$\delta n(E) = -\frac{\mathrm{Im}}{\pi} \lim_{\eta \to 0} [G_{11}^{*}(E+i\eta) - G_{11}(E+i\eta)]$$

$$= -\frac{\mathrm{Im}}{\pi} \lim_{\eta \to 0} \delta G_{11}(E+i\eta)$$

where $G_{11}^{*}(z)$ is characterized by the coefficients (a_n^{*}, b_n^{*2}) of its continued fraction expansion.

With the variations $\delta a_n = a_n^* - a_n$ and $\delta b_n = b_n^* - b_n$ is associated an effective perturbation potential δV, given in the recursion basis by :

$$\{n|\delta V|n\} = \delta a_n$$

$$\{n|\delta V|n+1\} = \{n+1|\delta V|n\} = \delta b_n$$

Let us now consider the well known DYSON expansion :

$$G^* = G + G\delta VG + G\delta VG\delta VG + \ldots$$

To first order in δV, we get :

$$G^* \sim G + G\delta VG$$

i.e.

$$\delta G_{11} \sim \{1|G\ VG|1\}$$

Since the recursion basis is orthonormal and complete, we have

$$\delta G_{11} \sim \sum_{n,m} \{1|G|n\}\{n|\delta V|m\}\{m|G|1\}$$

And finally δG_{11} is expressed as :

$$\delta G_{11} \sim \sum_n G_{1n}^2 \delta a_n + 2\sum_n G_{1n} G_{1n+1} \delta b_n \qquad (1)$$

Practically, the G_{1n} are calculated either by a decimation technique, as described by AOKI [8], or more simply through the convenient recurrence relation (see Appendix A) :

$$G_{1n} = b_{n-1} G_{nn}^{>} G_{1n-1} \qquad (2)$$

where

$$G_{nn}^{>}(z) = \frac{1}{z-a_n} - \frac{b_n^2}{z-a_{n+1}} - \ldots$$

The expression (1) will be of practical interest if only a few terms in the sum can be retained to restitute δG_{11}, i.e. if δa_n and δb_n are small enough when n tends to infinity.

2.2 Properties of the LGM :

To discuss the accuracy of the LGM, we first see how it approximates the differences $\delta\mu_n$ between the moments of the densities of states. If we suppose $\delta a_1 = 0$, i.e. $\delta\mu_1 = 0$, which is frequently the case, the expansion δG_{11} in powers of z^{-1} when z goes to infinity is given, from (1), by

$$\lim_{z\to\infty} \delta G_{11} = \frac{2b_1\delta b_1}{z_3} + \frac{2a_2 b_1\ \delta b_1 + b_1^2\ \delta a_2}{z_4} + 0\ (z^{-5})$$

since from relation (2) :

$$\lim_{z\to\infty} G_{1n} = \frac{b_1 \cdots b_{n-1}}{z_n} + 0\ (z^{-(n+1)})$$

Furthermore, because of the relation between the moments of the DOS and the chain parameters (see Appendix B) we get :

$$\delta\mu_2 = 2b_1\ \delta b_1 + (\delta b_1)^2$$

$$\delta\mu_3 = 2a_2 b_1\ \delta b_1 + b_1^2\ \delta a_2 + 2b_1\ \delta a_2 \delta b_1 + \delta a_2\ (\delta b_1)^2 + a_2\ (\delta b_1)^2$$

On the other hand the exact LAURENT expansion of $\delta G_{11}(z)$ is given by :

$$\lim_{z \to \infty} \delta G_{11}(z) = \sum_{n=0}^{\infty} \delta\mu_n . z^{-n-1}$$

So, the direct comparison between the two expansions shows that the neglected terms are of the form :

$$\frac{(\delta b_1)^2}{z^3} + \frac{2b_1\ \delta a_2\ \delta b_1 + a_2\ (\delta b_1)^2 + \delta a_2\ (\delta b_1)^2}{z^4} + 0\ (z^{-5})$$

Therefore, provided that the δa_n and δb_n are smaller than a typical energy, say for example the half bandwidth W of the reference DOS, we conclude that the corrective terms are negligible.

Before considering the convergence properties of the LGM, let us examine its application to the calculation of the differences in band energies $\delta E(\varepsilon)$. First let us define what is the formal expression for such a difference $\delta E(\varepsilon)$. If the atomic level is taken as the zero of energy, we have :

$$E(E_F) = \int^{E_F} t\ n(t)\ dt \text{ and } E(E_F^*) = \int^{E_F^*} t\ n^*\ (t)\ dt$$

where E_F and E_F^* are the FERMI levels, with the additional condition that the comparison is made at constant electronic charge N, i.e. at constant band filling :

$$\int^{E_F} n(t)dt = \int^{E_F^*} n^*(t)dt$$

If the variations of $\delta n(t)$ and $\delta E_F = E_F^* - E_F$ are small, to first order in δE_F, we get :

$$\delta E(E_{Fm}) \sim \delta E_F . E_{Fm} . n(E_{Fm}) + \int^{E_{Fm}} t\ \delta n(t)dt$$

where $E_{F_m} = E_F^*$ or E_F.

Similarly, the condition that the comparison is undertaken at constant filling N of the band yields :

$$0 \sim \delta E_F . n\ (E_{Fm}) + \int^{E_{Fm}} \delta n(t)\ dt$$

Therefore we get [9]:

$$\delta E(E_{Fm}) \sim \int^{E_{Fm}} dt\ (t - E_{Fm})\ \delta n(t)$$

or in an equivalent form :

$$\delta E(E_{Fm}) \sim -\int^{E_{Fm}} d\varepsilon \int^{\varepsilon} dt\ \delta n(t) \qquad (3)$$

$$\sim -\int^{E_{Fm}} d\varepsilon\ \delta N(\varepsilon)$$

If we are interested in the moments of $\delta E(\varepsilon)$, i.e. :

$$\delta m_n = \int t^n\ \delta E(t)dt$$

the following results is deduced :

$$\delta m_n = -\frac{1}{(n+1)\,(n+2)} \cdot \delta\mu_{n+2} \tag{4}$$

This important relation between the moments of the band energy difference and those of the DOS difference is a direct consequence of a general theorem about the properties of a function, the first moments of which are known :

Let f(t) denote a real continued function in nearly all the interval [a,b] such that $\int_a^b t^k f(t)\,dt = 0$ for k = 0,1, ..., n-1

Then, if f is not the identically null function, f(t) has at least n roots in]a,b[[9-11].

This theorem will be of particular interest in the following sections. In the present case it means that if n(E) and $n^*(E)$ have the same n first moments $\mu_0, \mu_1, \dots \mu_{n-1}$, $\delta n(N)$ and $\delta E(N)$ have at least n and (n-2) zeros respectively between N = 0 and 1 [9,10].

Using equations (1) and (3) the final expression of $\delta E(\varepsilon)$ is given by:

$$\delta E(\varepsilon) = \sum_n \phi_n(\varepsilon)\delta a_n + 2\sum_n \Psi_n(\varepsilon)\delta b_n$$

where
$$\phi_n(\varepsilon) = -\frac{\mathrm{Im}}{\pi}\lim_{\eta\to 0}\int^{\varepsilon} dt\,(t-\varepsilon)\; G_{1n}^2(t+i\eta) \tag{5}$$

$$\Psi_n(\varepsilon) = -\frac{\mathrm{Im}}{\pi}\lim_{\eta\to 0}\int^{\varepsilon} dt\,(t-\varepsilon)\,G_{1n}(t+i\eta)\,G_{1n+1}(t+i\eta)$$

Assuming n(E) to be the typical DOS of an average medium, the quantities ϕ_n and Ψ_n are calculated from the G_{1n} which characterize this medium. If we compare two systems, the average medium only acts through these integrated quantities ϕ_n and Ψ_n, so that we can predict that the difference δE will not be very sensitive to the choice of this medium.

To conclude the present section let us examine the convergence properties of the LGM. In order to get an analytical formulation of this problem, we first study the simple case of the constant semilinear chain considered here as our reference medium. For this example, $a_n = a$, $b_n = b$ for all n and the DOS is given by a semi-elliptic function with band edges a ± 2b. $G^>_{nn}$ is simply expressed as :

$$G^>_{nn} = \frac{e^{i\varphi}}{n}\;\forall n$$

where $\varphi = \mathrm{Arc\,cos}\;(\frac{E-a}{2b})$

From formula (2) we get : $G_{1n} = \frac{e^{in\varphi}}{b}\;\forall n$

The fluctuations of DOS with respect to the semi-elliptic DOS :

$$n(E) = \frac{\sin\varphi}{\pi b^2}$$

are given, according to (1), by :

$$\delta n(E) = \frac{1}{\pi b^2}\left[\sum_n \delta a_n \sin 2n\varphi + 2\sum_n \delta b_n \sin(2n+1)\varphi\right]$$

This relation illustrates for example the influence of internal singularities (if any), at least when the condition δa_n, $\delta b_n \ll 2b$ is fulfilled, so that the changes in the recursion chain parameters are just the FOURIER coefficients of δn (E), i.e. :

$$\delta a_n = 2b^2 \int_0^{\Pi} \delta n\,(E)\ \sin 2n\varphi\ d\varphi$$

$$\delta b_n = 2b^2 \int_0^{\Pi} \delta n\,(E)\ \sin (2n+1)\varphi\ d\varphi$$

a result which was previously established by HAYDOCK [7].

The ϕ_p and Ψ_p are given by integrating the TCHEBYSHEV polynomials of second kind :

$$\Phi_p(E) = \Lambda_{2p}(E) \qquad (6)$$

$$\psi_p(E) = \Lambda_{2p+1}(E) \qquad \text{where}$$

$$\Lambda q(E) = \Pi^{-1} \left[\frac{2\sin q\varphi}{(q-1)(q+2)} - \frac{\sin(q-2)\varphi}{(q-2)(q-1)} - \frac{\sin(q+2)\varphi}{(q+1)(q+2)} \right]$$

and the band filling is expressed as :

$$N(E) = \frac{1}{\Pi}(\Pi - \varphi) + \frac{\sin 2\varphi}{2\Pi}$$

When n goes to infinity, the integrated quantities ϕ_n and Ψ_n behave as $\sin 2n\varphi/n^2$ while the G_{1n} are independent of n. Furthermore if we plot ϕ_n and Ψ_n as a function of the band filling N, these functions, for fixed n, display (2n-3) and (2n-2) zeros respectively within the edges, as a consequence of the above-mentioned general theorem (see Fig. 1).

In the general case, we deduce from this particular model weight function that, as far as the chain parameters approach their asymptotic limits, G_{1n} behave as $f(E)\ e^{in\varphi(E)}$, where $E = a_\infty + 2b_\infty \cos\varphi$, and ϕ_n, Ψ_n are proportional to $\frac{\sin n\varphi(E)}{n^2}$ (with $2b_\infty = W$).

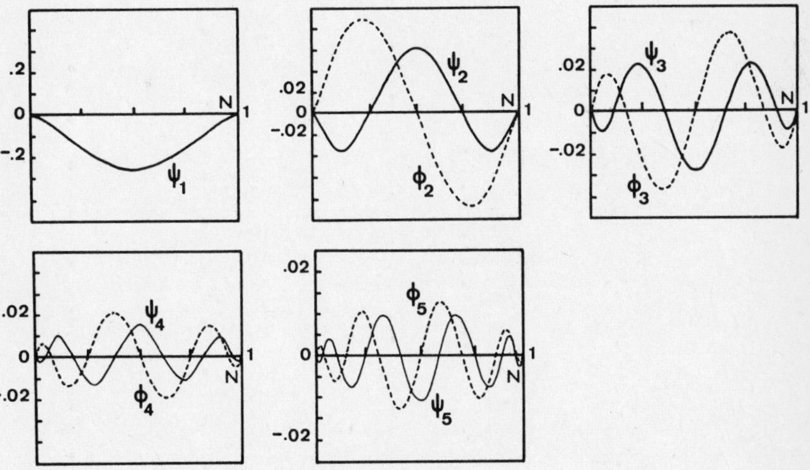

Fig. 1 – Plot of the first ϕ_n and ψ_n given by relations (6) versus band filling N for the semielliptic DOS as the reference medium, with a = 0 and b = 1.

Thus from the fact that ϕ_n and Ψ_n vary as n^{-2} and present an increasing number of zeros when n increases, we conclude that the analytical expansion (5) of $\delta E(\varepsilon)$ converges more rapidly than the corresponding expansion of $\delta n(\varepsilon)$. The LGM is therefore a suitable method to calculate in a simple way differences in band energies, with the additional advantage that the specific contribution of δa_n and δb_n, and therefore of $\delta\mu_{2n-1}$ and $\delta\mu_{2n}$, to the difference δE are directly apparent, which is not the case when one calculates differences between two continued fractions.

One can notice that a partial treatment of the LGM was given by HAYDOCK [7] in terms of orthogonal polynomials and by MAGNUS [12] from an analysis of the error induced by a modification of the continued fraction coefficients.

3. Application of the LGM to the determination of the stability of pure transition metals

We assume that the main contribution to the cohesive energy of a pure transition metal comes principally from the d band [2,10,13].

Taking the d atomic energy level as the origin of energy, the cohesive energy may be written as :

$$E_c = -E_b = -\int^{E_F} En(E)dE$$

where E_b is the band energy, E_F the FERMI level and n (E) the d DOS. The d electrons are described within the well – known tight binding hamiltonian:

$$H = \sum_{n\lambda} |n\lambda\rangle\, \varepsilon_n \,\langle n\lambda| + \sum_{\substack{nm\\ n\neq m\\ \lambda\mu}} |n\lambda\rangle\, \beta^{\lambda\mu}_{nm} \,\langle m\mu|$$

where n, m are lattice sites, λ and μ are orbital labels ($\lambda,\mu = 1,5$) and ε_n is the atomic energy level at site n, i.e. we neglect crystalline field effects. For pure metals $\varepsilon_n = 0$. The hopping integrals $\beta^{\lambda\mu}_{nm}$ are expressed in the simplest scheme by the three SLATER-KOSTER parameters [14] : $dd\sigma$, $dd\pi$, $dd\delta$. When comparing different topologically close packed structures (e.g. FCC, BCC, HCP, A15 and LAVES phases) we also need a phenomenological law of variation with distance of these parameters. Here we assume that :

$$\beta(R) = \beta(Ro)\, e^{-qR_o\left[\frac{R}{R_o}-1\right]}$$

with $\beta = dd\sigma, dd\pi, dd\delta$ and $qRo = 3$ [10,15]

$dd\sigma\,(Ro) = -2$
$dd\pi\,(Ro) = 1$
$dd\delta\,(Ro) = 0$ where by convention Ro is the smallest distance in the A15 structure [15].

For a geometrical description of the structures, we refer to SINHA [16] (see also [17]).

To apply the LGM, we have first to estimate the order of magnitude of $\delta a_n/b_\infty$ and $\delta b_n/b_\infty$. Table 1 shows that up to n = 2-3 the set of continued fraction coefficients associated with different close-packed structures, with the supplementary condition that all these phases are examined <u>at constant volume</u>, are quite similar, while the a_∞, b_∞ are also almost identical.

Table 1

	FCC	HCP	BCC	A15	C14	C15	C36
$b_1^2=\mu_2$	6.906	6.906	7.073	7.216	7.585	7.585	7.585
$\mu_3/\mu_2^{3/2}$	-0.189	-0.189	-0.191	-0.200	-0.206	-0.206	-0.206
μ_4/μ_2^2	1.913	1.843	1.849	1.727	1.791	1.835	1.813
$\mu_5/\mu_2^{5/2}$	-1.246	-1.065	-0.750	-0.588	-0.688	-0.659	-0.663
μ_6/μ_2^3	5.297	4.544	4.590	3.600	3.916	4.226	4.071
a_2	-0.498	-0.498	-0.507	-0.537	-0.569	-0.569	-0.569
b_2^2	6.058	5.575	5.747	4.960	5.677	6.010	5.843
a_3	-1.581	-1.215	-0.168	0.371	0.233	0.312	0.274
b_3^2	8.278	6.568	9.425	6.019	6.629	7.749	7.212
a_∞	-1.385	-1.220	-1.037	-0.050	-0.425	-0.525	-0.475
b_∞^2	7.673	6.840	8.331	6.126	8.051	8.629	8.338

We have now to define the reference medium : an ideal case would be to take a topologically disordered medium, i.e. an amorphous state. In the present case, since the $\emptyset_n$ and Ψ_n are not so sensitive to the details of the average medium DOS, we define its coefficients (a_n, b_n^2) from the average moments calculated from those of the FCC-BCC-A15 and C36 structures (these structures have been chosen because they exhibit typical local environments). The results, with our choice of electronic parameters ($dd\sigma$, $dd\pi$, $dd\delta$ and qRo) are given in Table 2.

Table 2

a_1 = 0.00	a_2 = -0.53	a_3 = -0.30	a_∞ = -0.71
b_1 = 7.20	b_2 = 5.70	b_3 = 8.40	b_∞^2 = 7.60

Incidentally, the DOS corresponding to this set of coefficients has all the characteristics of that of an amorphous metal as obtained from a hard sphere model [18], i.e. we notice the existence of a pseudogap and of an asymmetrical shape (see Fig. 2).

The figure 3 shows the corresponding $\emptyset_n$ (N) and ψ_n (N) (see relations (5)) where N is the d band filling corresponding to the DOS of figure 2 (normalized to unity). We have a confirmation of the insensitivity of the $\emptyset_n$ and ψ_n to the choice of the average medium (see figures 1 and 3) but in the present case, since $\mu_{2n+1} \neq 0$, these functions are somewhat asymmetric.

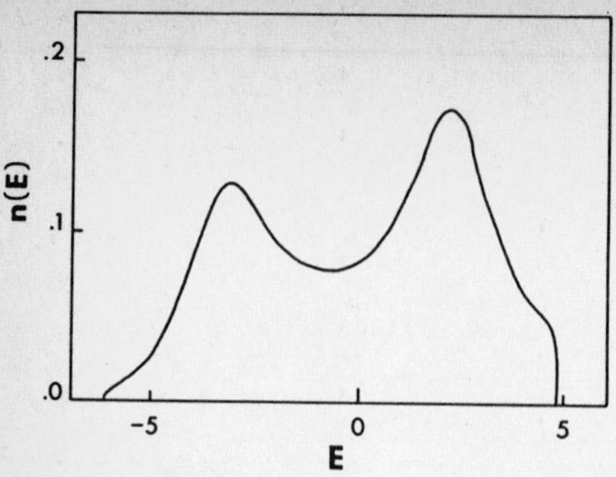

Fig. 2 — DOS of the average reference medium defined by the set of coefficients given in Table 2.

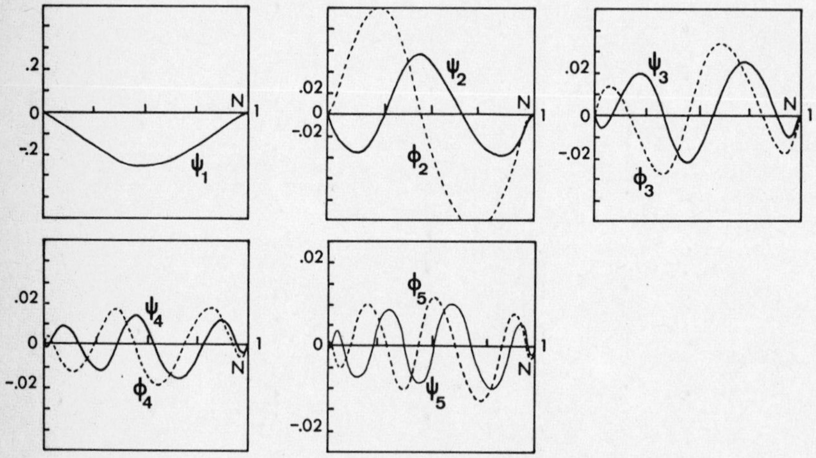

Fig. 3 — Plot of the first ϕ_n and ψ_n versus d band filling, corresponding to the DOS of Fig. 2.

From the general functions it is now easy to reconstruct a difference in band energies, i.e. a stability energy between two structures X and Y :

$$\Delta E^{(p-q)}_{X-Y} = \sum_{n=1}^{p} \Phi_n \delta a_n + 2 \sum_{m=1}^{q} \Psi_m \delta b_m \qquad (7)$$

As an example, consider the difference between FCC and BCC at constant volume. The δa_n, δb_n are given in table 3 and the result illustrated in figure 4 for different values of the couple (p, q) (see (7)).

Table 3 The first $\delta a_n = a_n$ (FCC)-a_n(BCC) and $\delta b_n = b_n$ (FCC) - b_n (BCC) calculated from the recursion method and corresponding to the electronic parameters given in the present section.

b_1 = -0.32	b_3 = -0.193	b_5 = -0.256
a_2 = -0.009	a_4 = 0.025	a_6 = -0.335
b_2 = 0.064	b_4 = -0.094	b_6 = 0.136
a_3 = -1.413	a_5 = -0.298	a_7 = -0.393

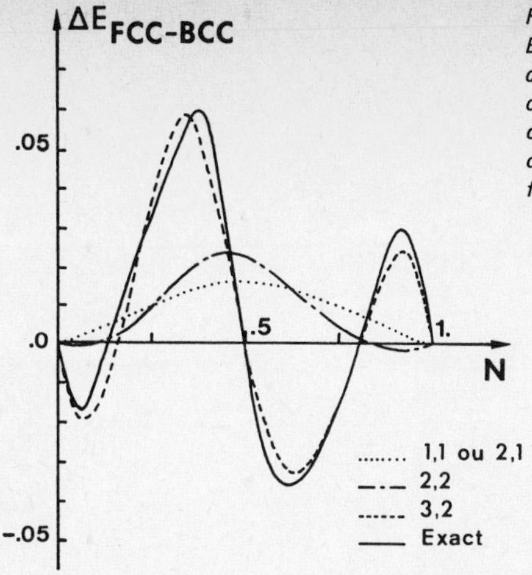

Fig. 4 – Stability energy between FCC and BCC as a function of the d band filling : the curves are labelled by the couples (p, q) defined in relation (7). The exact curve is obtained from a direct integration on the densities of states and 12 levels of continued fraction. (FCC is more stable than BCC when $\Delta E_{FCC-BCC} < 0$).

As also deduced from table 1, in general all close-packed structures principally differ from their μ_5, the comparison between FCC and BCC being the most typical (see table 3). Unfortunately, this important result is not easily understood. Indeed, from the definition of the nth moment of the DOS :

$$\mu_n = \frac{1}{5N_a} \sum_{\substack{i,j,k,..\\ \lambda,\mu,\nu,..}} \beta_{ij}^{\lambda\mu} \beta_{jk}^{\mu\nu} \cdots \beta_{li}^{\Pi\lambda}$$

where N_a is the number of lattice sites. μ_n represents a natural tool to relate the electronic structure to the topology, since it is obtained by calculating products of hopping integrals $\beta_{nm}^{\lambda\mu}$ associated with definite closed paths.

Up to μ_4 the situation is quite clear : the number of paths and their contributions lead to the fact that e.g. for μ_4, the self-retracing paths are of major importance and a simple calculation gives $\mu_4/\mu_2^2 \sim 2-1/Z$ where Z is the coordination number. Given that for all close-packed structures Z is between 12 and 14, it is quite obvious that the normalized μ_4 is not very sensitive to the topology. On the contrary, the situation is more complicated for μ_5 since the self-avoiding paths which are constructed on polygonal circuits are of great importance and very sensitive to the choice of the SLATER-KOSTER parameters. This discussion shows that fifth order moments are very important if one wants to describe differences between close-packed structures ; such effects have been underestimated previously [9,19].

To conclude this section, the LGM can be successfully applied to the calculation of band energy differences of pure transition metals, its accuracy only depending on the choice of the average medium. For example, it has been applied recently to the study of stacking faults in the HCP structure with introduction of a relaxation process ; in this case the method converges rapidly, since we know that $\delta a_\infty = \delta b_\infty = 0$, and the iterative calculation of the energy and the search of its minimum become quite easy [20]. This argument would apply to the study of any local defect.

4. Application of the LGM to the study of the stability properties of ordered transition metal alloys :

To simplify the problem, we restrict our study to the case of coherent ordered structures, i.e. to structures built up on the same underlying lattice. We first examine the moments of the DOS relative to a given configuration of an alloy $A_{1-c}\, B_c$ whose atomic energy levels are ε_A or ε_B depending on the chemical occupancy of each site. The average energy $\overline{\varepsilon} = (1-c)\ \varepsilon_A\ + c\ \varepsilon_B$ is chosen as the zero of energy and the alloy parameters are just the concentration c and what is commonly called the diagonal disorder $\delta = \varepsilon_B - \varepsilon_A$.

The hamiltonian H of the system is expressed as :

$$H = H_o + V$$

with
$$H_o = \sum_{\substack{nm\\ \lambda\mu \\ n\neq m}} |n\lambda>\beta_{nm}^{\lambda\mu} < m\mu| + \overline{\varepsilon} \sum_{n\lambda} |n\lambda><n\lambda|$$

$$V = \sum_n V_n = \sum_{n\lambda} (p_n - c)\delta\ |n\lambda> <n\lambda|$$

where p_n is the occupation number defined as :

$$p_n = \begin{cases} 1 & \text{if the site n is occupied by a B atom} \\ 0 & \text{otherwise.} \end{cases}$$

with the property : $\frac{1}{N_a} \sum_n p_n = c$

Since $\overline{\varepsilon} = 0$, H_o is just the hamiltonian of the pure metal. If we set :

$$\mu_n = \frac{1}{5N_a} \mathrm{Tr}\ H^n = \frac{1}{5N_a} \sum_{n\lambda} < n\lambda|\ H^n\ |\ n\,\lambda> \text{ and } \mu_n^o = \frac{1}{5N_a} \mathrm{Tr} H_0^n$$

the first moments of the alloy are given by :

$$\mu_o = 1$$

$$\mu_1 = \mu_1^o = 0 \text{ (because } \overline{\varepsilon} = 0)$$

$$\mu_2 = \mu_2^o + c\ (1-c)\ \delta^2$$

$$\mu_3 = \mu_3 + \frac{3}{Na} \sum_n (p_n - c)\delta . \frac{1}{5} \sum_\lambda <n\lambda|\ H_o^2\ |n\,\lambda> + c\ (1-c)\ (1-2c)\delta^3$$

If all the lattice sites have the same environment, which is the case for FCC, HCP, BCC,... $\sum_\lambda < n\lambda|H_o^2|n\lambda>$ is independent of n and thus the associated term does not contribute to μ_3.

Therefore, when all the sites are geometrically equivalent, the moments up to μ_3 are independent of the chemical configuration.

$$\mu_4 = \mu_4^o + 4c\ (1-c)\delta^2\mu_2^o\ + c(1-c)(1-3c+3c^2)\ \delta^4 + \frac{2}{5N_a} \sum_{\substack{nm\\ \lambda\mu \\ n\neq m}} (\beta_{nm}^{\lambda\mu})^2\ (p_n - c)\ (p_m - c)\delta^2$$

If we define $\delta\mu_4$ by :

$$\delta\mu_4 = \mu_4 - <\mu_4>$$

where $\langle\mu_4\rangle$ denotes the statistical average of μ_4 in the completely disordered state (i.e. $\langle p_n p_m\rangle = c^2$ if $n \neq m$ and c if $n = m$) then:

$$\delta\mu_4 = \frac{2}{N_a} \sum_{\substack{n\,m \\ n\neq m}} \beta^2(R_{nm}) \; (p_n-c)(p_m-c)\delta^2$$

where $\beta^2(R_{nm}) = \frac{1}{5} \sum_{\lambda\mu} (\beta_{nm}^{\lambda\mu})^2$

In particular, if the underlying lattice is a FCC one, twelve first neighbours are retained and $\beta^2(R) = \overset{o}{\mu}_2/12$. Therefore $\delta\mu_4 < 0$ (> 0) if $(p_n-c)(p_m-c) < 0$ (> 0) which is the case of ordering (segregation).

In the same way, we obtained for μ_5 :

$$\mu_5 = \langle\mu_5\rangle + \delta\mu_5$$

where $\langle\mu_5\rangle = \overset{o}{\mu}_5 + 5c(1-c)\delta^2\overset{o}{\mu}_3 + 5c(1-c)(1-2c)\delta^3\overset{o}{\mu}_2 + c(1-c)(1-2c)(1-2c+2c^2)\delta^5$

$$\delta\mu_5 = \frac{5}{N_a} \sum_{\substack{n\,m \\ n\neq m}} (p_n-c)(p_m-c)\,\delta^2 \cdot \frac{1}{5} \sum_{\lambda\mu} \langle n\lambda|H_o^2|m\mu\rangle \langle m\mu|H_o|n\lambda\rangle + \frac{5}{N_a} \sum_{\substack{n\,m \\ n\neq m}} (p_n-c)^2(p_m-c)\delta^3\,\beta^2(R_{nm})$$

If the self-retracing path is schematically drawn as a loop, the diagrammatic representations of the two terms of $\delta\mu_5$ are respectively given by :

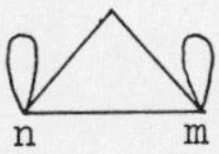

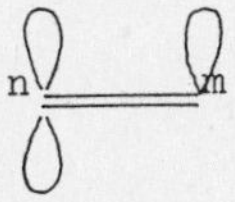

and one deduces that μ_5 only depends on the chemical occupation (via p_n and p_m) of neighbouring sites reached in one hop.

μ_6 is more complicated, but extrapolating the previous results, $\delta\mu_6$ will now contain contributions involving chemical occupation of two sites

separated by two hops : 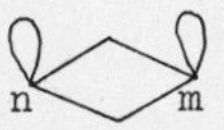 , 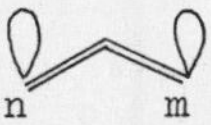 and also contributions involving three neighbouring sites : 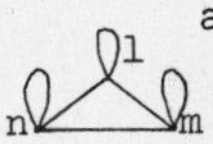 proportional to $(p_n-c)(p_m-c)(p_l-c)\,\delta^3\,\overset{o}{\mu}_3$. In general $\overset{o}{\mu}_3$ is small (see table 1) and these terms are negligible.

Then we conclude that ordering effects appear first in μ_4. The corresponding contribution is proportional to δ^2 and only involves near neighbours directly connected by non-vanishing hopping integrals (e.g. first neighbours for the FCC and HCP lattices ; first and second neighbours for the BCC one).

If one remembers the general theorem given in section 2 (§2.2) the difference δE between the energies of two coherent ordered structures as a function of the average band filling N should behave as indicated in Fig. 5.

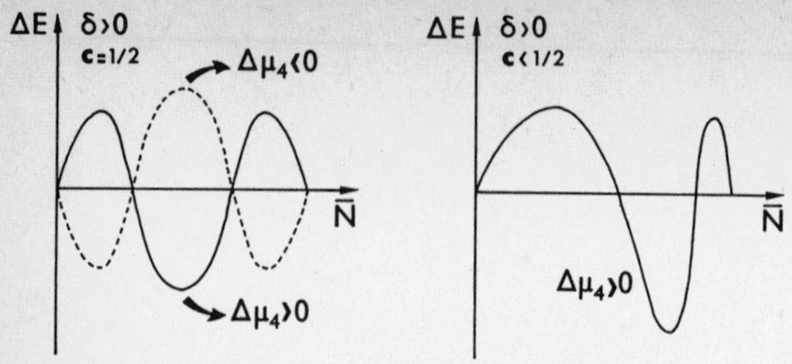

Fig. 5 – Schematic variations of the stability energy relative to two coherent ordered structures as a function of the average d band filling $\bar{N}$, for $\delta > 0$ and for $c = 1/2$ or $c < 1/2$.

Note that if we compare two ordered structures having the same near neighbour occupancy, we have to consider μ_6 and ordering effects will be driven by the pairs whose sites are reached in two jumps, e.g. the 2nd, the 3rd and the 4th neighbours for a FCC lattice [6].

From this detailed moment analysis, the ordering energy for a given ordered structure can be expressed to a first approximation as a sum of effective pair interactions :

$$\delta E_{ord} \simeq \frac{1}{N_a} \sum_{\substack{n\,m \\ n \neq m}} (p_n - c)(p_m - c) V_{nm}$$

If we suppose these interactions to be roughly determined by their first non zero moments, we have :

$$\delta E_{ord} = 2\,\Psi_2\,\delta b_2 + \Phi_3\,\delta a_3$$

because $\delta\mu_1 = \delta\mu_2 = \delta\mu_3 = 0$ and therefore $\delta a_1 = \delta b_1 = \delta a_2 = 0$

The relation between moments and continued fraction coefficients (see Appendix B) gives in a straightforward manner :

$$\delta b_2 = \frac{\delta\mu_4}{b_1^2 b_2}, \quad \delta a_3 = \frac{\delta\mu_5}{b_1^2\, b_2^2} - (2a_2 + a_3)\,\frac{\delta\mu_4}{b_1^2\, b_2^2}$$

where $\delta\mu_4$ and $\delta\mu_5$ were explicited above.

If the contribution of $\langle n\lambda|H_0^2|m\mu\rangle \langle m\mu|H_0|n\lambda\rangle$ to $\delta\mu_5$ is small, we finally express the effective pair interaction between sites n and m separated by one hop as :

$$V_{nm} \equiv V(R_{nm}) = \left[\frac{2\delta^2 b_2 \Psi_2 - 2\delta^2(2a_2 + a_3)\,\Phi_3 + 5(1-2c)\,\delta^3\Phi_3}{5 b_1^2\, b_2^2}\right] \sum_{\lambda\,\mu} (\beta_{nm}^{\lambda\mu})^2 \qquad (7)$$

with, in practice :

$$\sum_{\lambda\,\mu} (\beta_{nm}^{\lambda\mu})^2 = dd\sigma^2(R_{nm}) + 2dd\Pi^2(R_{nm}) + 2dd\delta^2(R_{nm})$$

As far as $\emptyset_n$ and Ψ_n are those of the pure metal (see § 3) it appears from relation (7) that our development is valid if $\delta/W \ll 1$, i.e. in the limit of a weak diagonal disorder. Moreover, these pair interactions are

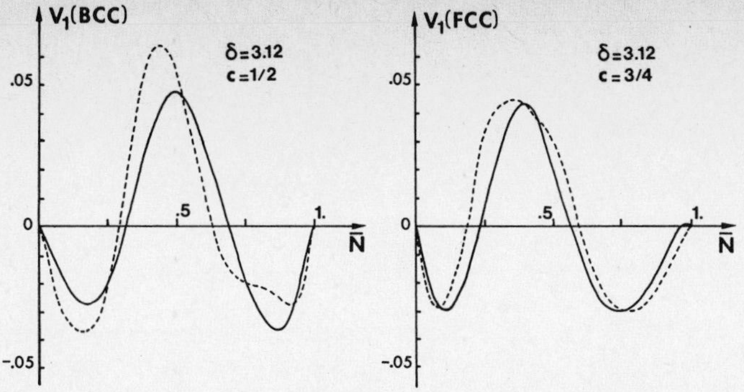

Fig. 6 – Nearest neighbour effective pair interaction V_1 relative to BCC and FCC structures : —— V_1 given by relation (7) ; – – – V_1 deduced from a GPM calculation ($R_1^{BCC} = 0.545$, $R_1^{FCC} = 0.561$ in unit of the lattice parameter of the A 15 structure [15]).

quite universal and the relation (7) gives us an analytical expression of the V_{nm} in term of V ($\overline{N}$, c, δ ,R_{nm}). This result can be compared with that obtained using a more quantitative theory, the Generalized Pertubation Method based on a CPA description of the disordered state [3,4,15]. (see Figure 6).

In the case of strong disorder, i.e. $\delta/W \gtrsim 1$, formula (7) breaks down but the $\emptyset_n$ and Ψ_n can be renormalized by using the CPA medium as a reference, and finally it is possible to discuss the stability of transition alloys for different states of order and different crystalline structures [21].

5. Conclusion

Apart from the practical aspect of the LGM which leads to a precise evaluation of band energies, this method yields a simple tool to describe the electronic properties of transition metals and alloys and to investigate the origin of their relative phase stabilities. Although we have mainly discussed the case of pure metals and coherent ordered alloys, the LGM can be extended to the case of non-coherent ordered structures [21] and transition compounds such as transition carbides and nitrides [22].

Acknowledgments The authors are indebted to G. Treglia for several fruitful discussions.

Appendix A

The expression of G_{1n} is simply given by :

$$G_{1n} = b_1 b_2 \ldots b_{n-1} \, G_{11}^{>} \, G_{22}^{>} \ldots G_{nn}^{>}$$

where

$$G_{nn}^{>}(z) = \frac{1|}{|z-a_n} - \frac{b_n^2|}{|z-a_{n+1}} - \ldots$$

We have indeed :

$$G^{>}_{nn}(z) = \frac{1}{z-a_n-b_n^2\, G^{>}_{n+1n+1}(z)} \quad \text{with } G^{>}_{11}(z) = G_{11}(z)$$

Let us vary a_n ; to first order we have :

$$\delta G^{>}_{11} = b_1^2\, G_{11}^2\, \delta G^{>}_{22}$$

$$\delta G^{>}_{22} = b_2^2\, G^{>2}_{22}\, \delta G^{>}_{33}$$

$$\delta G^{>}_{nn} = \delta a_n\, G^{>2}_{nn}$$

so that :

$$\delta G_{11} = b_1^2 b_2^2 \ldots b_{n-1}^2 G^{>2}_{11} G^{>2}_{22} \ldots G^{>2}_{nn}\, \delta a_n.$$

Since we have also :

$$\delta G_{11} = G_{1n}^2\, \delta a_n \text{ (see relation (1) in § 2.1)}$$

the result follows.

Appendix B

The LAURENT expansion of G(z) :

$$G(z) = \sum_{n=0}^{\infty} \mu_n\, z^{-(n+1)}$$

and its JACOBI continued fraction development :

$$G(z) = \frac{1 |}{|z-a_1} - \frac{b_1^2 |}{|z-a_2} - \ldots.$$

lead, by a direct identification of the successive power of z^{-1}, to a relation between the continued fraction coefficients and the moments of the DOS :

$$a_n = f(\mu_0, \mu_1, \ldots, \mu_{2n-1})$$

$$b_n = f(\mu_0, \mu_1, \ldots, \mu_{2n}) \quad \text{where } \mu_n = \frac{1}{N_a} \mathrm{Tr} H^n$$

As a consequence of the invariance property of the trace with respect to the basis, the first moments, in the recursion basis, are merely expressed as :

$$\mu_0 = 1$$

$$\mu_1 = a_1$$

$$\mu_2 = a_1^2 + b_1^2$$

$$\mu_3 = a_1^3 + 2a_1 b_1^2 + a_2 b_1^2$$

$$\mu_4 = a_1^4 + 3a_1^2 b_1^2 + 2a_1 a_2 b_1^2 + a_2^2 b_1^2 + b_1^2 b_2^2 + b_1^4$$

1 R. HAYDOCK, V. HEINE and M.J. KELLY, J. Phys. C : Sol. St. Phys., 5, 2845 (1972) and 8, 2591 (1975).

2 D.G. PETTIFOR in : "Physical Metallurgy", eds R.W. CAHN and P. HAASEN (North Holland, Amsterdam and New York), chap. 3 (1984).

3 A. BIEBER and F. GAUTIER, Sol. St. Com., 38, 1219 (1981).

4 F. DUCASTELLE and F. GAUTIER, J. Phys. F. : Met. Phys., 6, 2039 (1976).

5 A. BIEBER and F. GAUTIER, Sol. St. Com., 39, 149 (1981).

6 A. BIEBER, F. DUCASTELLE, F. GAUTIER, G. TREGLIA and P. TURCHI, Sol. St. Com., 45, 585 (1983).

7 R. HAYDOCK, Sol. St. Phys., 35, 216-294 (1980).

8 H. AOKI, J. Phys. C: Sol. St. Phys., 13, 3369 (1980).

9 F. DUCASTELLE and F. CYROT-LACKMANN, J. Phys. Chem. Sol., 32, 285 (1971).

10 F. DUCASTELLE, Thesis, Orsay (1972).

11 J. DIEUDONNE, "Calcul Infinitésimal" (ed. HERMANN, Paris), 162 (1968).

12 A. MAGNUS, "Riccati acceleration of Jacobi continued fractions and Laguerre-Hahn orthogonal polynomials", Rapport n° 26 (Cabay, Louvain-la-Neuve) (1983).

13 J. FRIEDEL in : "The Physics of Metals", ed J.M. ZIMAN (Cambridge University Press, Londres), ch. 8 (1969).

14 J.C. SLATER and G. KOSTER, Phys. Rev., 94, 1498 (1954).

15 P. TURCHI, G. TREGLIA and F. DUCASTELLE, J. Phys. F. : Met. Phys., 13, 2543 (1983).

16 A.K. SINHA, Progr. Mater. Sci., 15 (1972).

17 P. TURCHI, Thesis, Paris VI (1984).

18 S.N. KHANNA and F. CYROT-LACKMANN, Phys. Rev., B21, 1412 (1980).

19 K. HIRAI and J. KANAMORI, J. Phys. Soc. Jap., 50, 2265 (1981).

20 B. LEGRAND, Thesis, Paris VI (1984).

21 F. DUCASTELLE, Mat. Res. Soc. Symp. Proc., ed. Th. TSAKALAKOS (North-Holland, Amsterdam and New York), 21, 375 (1984).

P. TURCHI and F. DUCASTELLE, to be published.

22 J.P. LANDESMAN, P. TURCHI, F. DUCASTELLE and G. TREGLIA, Mat. Res. Soc. Symp. Proc., ed. Th. TSAKALAKOS (North-Holland, Amsterdam and New York), 21, 363 (1984) and to be published.

Response Functions and Interatomic Forces

M.W. Finnis

Theoretical Physics Division, Building 424.4, A.E.R.E. Harwell,
Oxfordshire OX11 ORA, United Kingdom

D.G. Pettifor

Department of Mathematics, Imperial College of Science and Technology,
London SW7 2BZ, United Kingdom

If a tight-binding Hamiltonian H is perturbed by some displacement of the atoms, then the second order change in energy is given by :

$$\Delta U^{(2)} = \Delta U^{(2)}_{rep} - (1/\pi)\, Im\, Tr \sum_{\sigma} \int^{\varepsilon_F} \{G^{\sigma}(\varepsilon)\Delta H^{(2)} + \tfrac{1}{2} G^{\sigma}(\varepsilon)\Delta H^{(1)} G^{\sigma}(\varepsilon)\Delta H^{(1)}\} d\varepsilon,$$

where $G^{\sigma}(\varepsilon)$ is the σ–spin Green function for the unperturbed lattice, and $\Delta H^{(1)}$ and $\Delta H^{(2)}$ are the first and second order changes in the Hamiltonian whose matrix elements between sites and orbitals are obtained from the Slater–Koster two-centre integrals expanded as a Taylor series in the atomic displacements $u_{a\alpha}$. Force constants $\phi^{ab}_{\alpha\beta}$ are the coefficients of $-u_{a\alpha}u_{b\beta}$ in the above expansion of $\Delta U^{(2)}$. The recursion method is used to calculate the matrix elements of $G^{\sigma}(\varepsilon)$. The energy integrals of products of these are response functions, which should satisfy a sum rule over lattice sites to equal the density of states. By calculating the density of states directly and via the response function sum rule we have a useful test of the range and accuracy of the response functions which we have calculated to the tenth shell of neighbours in the bcc lattice ([333]a/2 and [115]a/2), using 13 – 15 exact levels of the continued fraction for the canonical d-band model. Results of this test are presented, followed by calculations of force constants. Features of the phonon spectra in bcc transition metals are explained.

1 Introduction

A theory of interatomic force constants in materials is essential for understanding their phonon spectra. This is now possible on the basis of a tight-binding (TB) description of the electronic structure[1–4]. The present method, first described in [3] , differs from the other approaches by working entirely in real space, making use of the recursion method. It could therefore be applied to arbitrary atomic configurations.

Within TB theory the total energy of a transition metal is written :

$$U = U_{rep} + \sum_{am\sigma} \int^{\varepsilon_F} \varepsilon n_a^{m,\sigma}(\varepsilon) d\varepsilon, \tag{1}$$

where $n_a^{m,\sigma}$ is the density of states for a given spin σ which is associated with orbital m on atom a. Henceforth for simplicity we shall drop the spin index, assuming spin degeneracy. For our force constant calculations (see below and [3]) we assumed that m runs over the nine s,p and d orbitals required to fit the first-principles transition-metal band structure of V [5] within the orthogonal two-centre TB approximation[6] . U_{rep} is an empirical short-range repulsive contribution which counters the attractive band term[7] . If the TB Hamiltonian H is perturbed by some displacement of the atoms, then the change in energy is given to second order by[8–9] :

$$\Delta U^{(2)} = \Delta U^{(2)}_{rep} - (1/\pi)\, Im\, Tr\, 2\int^{\varepsilon_F} \{G(\varepsilon)\Delta H^{(2)} + \tfrac{1}{2} G(\varepsilon)\Delta H^{(1)} G(\varepsilon)\Delta H^{(1)}\} d\varepsilon, \tag{2}$$

where G is the Green function for the unperturbed lattice, and $\Delta H^{(1)}$ and $\Delta H^{(2)}$ are the first– and second-order changes in the two-centre TB Hamiltonian.

The interatomic force constants $\phi^{ab}_{\alpha\beta}$ may be obtained by perturbing the lattice so that atom a moves in the α direction and atom b moves in the β direction ($\alpha,\beta=x,y,z$). It follows from (2) that :

$$\phi^{ab}_{\alpha\beta} = \hat{\phi}^{ab}_{\alpha\beta} + 2\sum_{dnq}(\delta\rho^{nq}_{bd,a\alpha}\nabla_\beta h^{qn}_{db} + \delta\rho^{nq}_{ad,b\beta}\nabla_\alpha h^{qn}_{da}), \tag{3}$$

where h^{qn}_{db} is the two-centre hopping integral linking orbitals n and q on atoms b and d respectively. The induced change in the bond charge (per spin) is given by :

$$\delta\rho^{nq}_{bd,a\alpha} = \sum_{cmp}(\chi^{np,mq}_{bc,ad} + \chi^{nm,pq}_{ba,cd})\nabla_\alpha h^{pm}_{ac}, \tag{4}$$

with the response function χ defined by :

$$\chi^{np,mq}_{bc,ad} = (1/\pi)\, Im \int^{\varepsilon_F} G^{np}_{bc}(\varepsilon)G^{mq}_{ad}(\varepsilon)d\varepsilon. \tag{5}$$

The contributions from $\Delta H^{(2)}$ to the trace in (2) are short-ranged because the hopping integrals only extend to the second shell of neighbours in the bcc structure, and they have been grouped together with $\Delta U^{(2)}_{rep}$ into the empirical term $\hat{\phi}^{ab}_{\alpha\beta}$, as in [1] .

Equation (3) omits any shift in the diagonal elements of the Hamiltonian with the change in atomic positions, which should be included in a more accurate theory in order to achieve consistency between the compressibility as calculated from the force constants (equivalent to the long wavelength limit of the longitudinal phonon frequency/wavevector) and as calculated by homogeneous dilatation (Finnis *et al.* to be published).

2 Matrix Elements

Let us now examine the expression for the force constants in more detail. It is necessary to calculate the matrix elements of ∇h and the elements of the Green function matrix G, which we consider in turn. In the full calculations reported in [3] , the overlap integrals fitted to the band structure of V [5] were used, but it is also instructive to consider some results obtained when hybridisation with s and p orbitals is neglected. The d–orbitals are labelled 1 to 5 and assigned symmetries xy, yz, zx, x^2-y^2, and $3z^2-r^2$ respectively. Within a canonical d–band description[10] the overlap integrals reduce to a simple form[11] depending only on the bandwidth. For the first and second neighbours in bcc there are then only three independent matrices of the derivatives of h which we may write as follows :

$$\left(\frac{\partial h}{\partial x}\right)_{<111>} = \frac{5\beta^{(1)}}{3\sqrt{3}}\begin{bmatrix} 2 & 10 & 0 & -7 & 5\sqrt{3} \\ 10 & 10 & 10 & 0 & 0 \\ 0 & 10 & 2 & -11 & \sqrt{3} \\ -7 & 0 & -11 & -5 & -2\sqrt{3} \\ 5\sqrt{3} & 0 & \sqrt{3} & -2\sqrt{3} & -9 \end{bmatrix} \tag{6}$$

$$\left(\frac{\partial h}{\partial x}\right)_{<002>} = 5\beta^{(2)}\begin{bmatrix} 0 & 1 & 0 & 0 & 0 \\ 1 & 0 & 0 & 0 & 0 \\ 0 & 0 & 0 & 1 & -2\sqrt{3} \\ 0 & 0 & 1 & 0 & 0 \\ 0 & 0 & -2\sqrt{3} & 0 & 0 \end{bmatrix} \tag{7}$$

$$\left(\frac{\partial h}{\partial z}\right)_{<002>} = 5\beta^{(2)} \begin{bmatrix} 1 & 0 & 0 & 0 & 0 \\ 0 & -4 & 0 & 0 & 0 \\ 0 & 0 & -4 & 0 & 0 \\ 0 & 0 & 0 & 1 & 0 \\ 0 & 0 & 0 & 0 & 6 \end{bmatrix} \quad \text{where :} \tag{8}$$

$$\beta = \frac{2W}{5}\left(\frac{S}{R}\right)^5 \frac{1}{R},$$

S = the Wigner–Seitz radius,
W = the Wigner–Seitz bandwidth,
R = the neighbour distance = $a\sqrt{3}/2$ for $\beta^{(1)}$ = a for $\beta^{(2)}$.

Thus in the canonical case the only free parameter is W, which we take as the unit of energy in the calculations. In the case of V, with the fitted band-structure parameters[5] , we retained only the $d-d$ overlaps, characterised by $dd\sigma$, $dd\pi$ and $dd\delta$, in the evaluation of ∇h. The matrix elements (6)–(8) are then linear combinations of these three overlap parameters.

3 Green Functions

We have exploited the cubic symmetry to reduce the number of Green function matrix elements (henceforth referred to as 'Green functions') and shorten the required computation times. With cubic symmetry we need only consider the matrix elements between a site at the origin and a single reference site in each shell of neighbours, including the origin itself. Symmetry of the Green function matrix between the real orbitals on two sites would leave 15 independent elements between the sites, but cubic symmetry further reduces this number. For example, the matrix between the origin site and itself is diagonal, with two independent elements :

$$G(000) = \begin{bmatrix} G_{11}^{(0)} & 0 & 0 & 0 & 0 \\ 0 & G_{11}^{(0)} & 0 & 0 & 0 \\ 0 & 0 & G_{11}^{(0)} & 0 & 0 \\ 0 & 0 & 0 & G_{44}^{(0)} & 0 \\ 0 & 0 & 0 & 0 & G_{44}^{(0)} \end{bmatrix}. \tag{9}$$

We introduce in (9) the notation $G_{pq}^{(n)}$ to mean the element of the Green function between orbital p at the origin and orbital q on the reference site in the n–th shell of neighbours. The Green functions to first and second neighbours each have four independent elements and take the following form :

$$G(111) = \begin{bmatrix} G_{11}^{(1)} & G_{12}^{(1)} & G_{12}^{(1)} & 0 & G_{15}^{(1)} \\ G_{12}^{(1)} & G_{11}^{(1)} & G_{12}^{(1)} & \frac{1}{2}\sqrt{3}G_{15}^{(1)} & -\frac{1}{2}G_{15}^{(1)} \\ G_{12}^{(1)} & G_{12}^{(1)} & G_{11}^{(1)} & -\frac{1}{2}\sqrt{3}G_{15}^{(1)} & -\frac{1}{2}G_{15}^{(1)} \\ 0 & \frac{1}{2}\sqrt{3}G_{15}^{(1)} & -\frac{1}{2}\sqrt{3}G_{15}^{(1)} & G_{44}^{(1)} & 0 \\ G_{15}^{(1)} & -\frac{1}{2}G_{15}^{(1)} & -\frac{1}{2}G_{15}^{(1)} & 0 & G_{44}^{(1)} \end{bmatrix}, \tag{10}$$

$$G(002) = \begin{bmatrix} G_{11}^{(2)} & 0 & 0 & 0 & 0 \\ 0 & G_{22}^{(2)} & 0 & 0 & 0 \\ 0 & 0 & G_{22}^{(2)} & 0 & 0 \\ 0 & 0 & 0 & G_{44}^{(2)} & 0 \\ 0 & 0 & 0 & 0 & G_{55}^{(2)} \end{bmatrix}. \tag{11}$$

The independent $G_{pq}^{(n)}$ were calculated using the recursion method[12] on a cluster of 14911 atoms, in which the central site is 15 second-neighbour hops from the surface. Starting with orbital p at the origin, the recursion method generates a and b coefficients for $G_{pp}^{(0)}$. To obtain $G_{pq}^{(n)}$ we made two recursion calculations in which the starting orbitals were symmetric and antisymmetric combinations of p at the origin and q at the reference site in shell n. Each continued fraction was terminated by an optimally chosen square root terminator[13] and tabulated for subsequent computations. Let us call the results $\langle p+q|G^{(n)}|p+q\rangle$ and $\langle p-q|G^{(n)}|p-q\rangle$. The required off-diagonal element of the Green function was then obtained by subtraction :

$$G_{pq}^{(n)} = \tfrac{1}{4}\left(\langle p+q|G^{(n)}|p+q\rangle - \langle p-q|G^{(n)}|p-q\rangle \right). \tag{12}$$

The canonical results out to second neighbours are shown Fig.1. Characteristically they are increasingly oscillatory functions of energy at further neighbour separations[14–16].

The physical significance of the diagonal elements of G in (9) is well known in relation to the local density of states. The total local density of states on the origin site, including both spins, is given by :

$$n_0(\varepsilon) = (-1/\pi)\, Im\, [6G_{11}^{(0)}(\varepsilon+i0) + 4G_{44}^{(0)}(\varepsilon+i0)]. \tag{13}$$

Figure 2 shows the results of calculating n_0 to various numbers of levels in the continued fraction.
The 'exact' result shown is a conventional k−space calculation[17] , which exhibits the van–Hove singularities. To obtain 20 exact levels of the continued fraction a cluster of 34481 atoms was generated; however, the result to 15 levels was deemed sufficiently accurate to use the smaller cluster for the other Green function calculations.

The physical significance of $G_{pq}^{(n)}$ lies in the interpretation of its integral as a bond order[18] as follows. Consider first the familiar result for the band energy :

$$\int^{\varepsilon_F} \varepsilon n_0(\varepsilon)d\varepsilon = -\,(2/\pi)\, Im \sum_p \int^{\varepsilon_F} \varepsilon G_{pp}^{(0)}(\varepsilon)d\varepsilon. \tag{14}$$

Replacing $Im\ \varepsilon G$ by $Im\ HG$ (since $G = (\varepsilon-H)^{-1}$) :

$$\int^{\varepsilon_F} \varepsilon n_0(\varepsilon)d\varepsilon = -(1/\pi)\, Im \sum_{pqa} \int^{\varepsilon_F} h_{0a}^{pq} G_{a0}^{qp}(\varepsilon)d\varepsilon. \tag{15}$$

The elements of G in (15) are given for the reference sites in each shell by (10) and (11). Only the central, first and second neighbour Green functions enter in our bcc model; that is the range of the Hamiltonian matrix elements h. For $h_{00} = 0$, (15) expresses the band energy as a sum over bonds of the hopping integrals weighted by a bond order. The positive and negative oscillations of $(-1/\pi)\ Im\ G_{pq}^{(n)}$ can therefore be identified with bonding and antibonding states, according to the sign of h_{pq}.

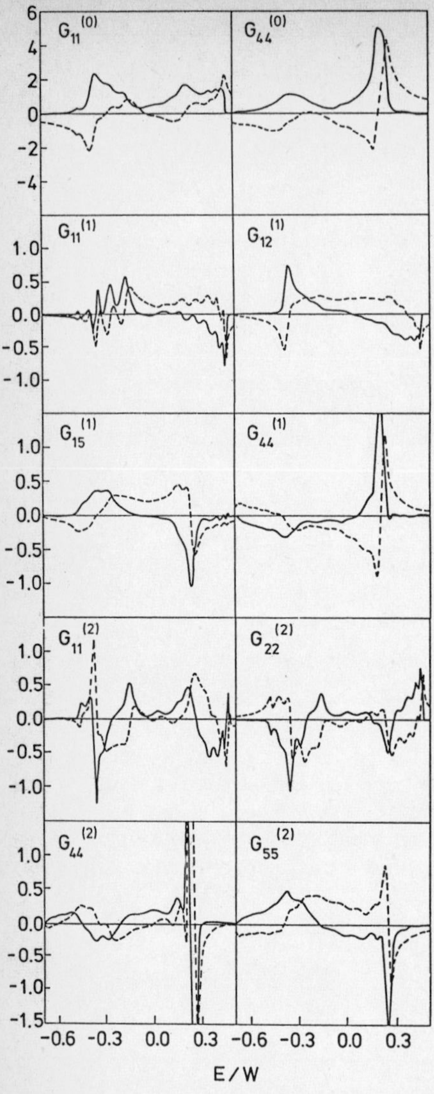

Figure 1 : The real (dotted curves) and imaginary (full curves) parts of the Green functions connecting the central site to itself and to its first and second neighbours. Real and imaginary parts have been scaled by $1/\pi$ and $-1/\pi$ respectively.

3 Response Functions

It is convenient to introduce a condensed notation for the independent response functions following our notation for the Green functions, thus :

$$\chi_{pq,rs}^{(m,n)}(\varepsilon_F) = (1/\pi)\, Im \int^{\varepsilon_F} G_{pq}^{(m)}(\varepsilon) G_{rs}^{(n)}(\varepsilon) d\varepsilon. \tag{16}$$

The largest in magnitude tend to be those involving a central site Green function and these are displayed in Fig.3.

For high accuracy the integrals in (16) were performed by Simpson's rule on a contour in the imaginary half-plane, but for the V force constant calculations we found good accuracy was achieved by the trapezoidal rule at 200 points along the real axis.

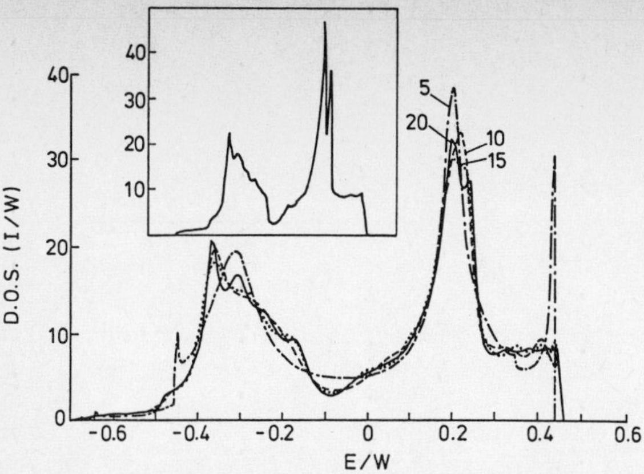

Figure 2 : The total density of states on the central site, including a factor 2 for spin degeneracy, calculated to 5, 10, 15 and 20 levels of the continued fraction. The insert shows the 'exact' result of MUNIZ[17] .

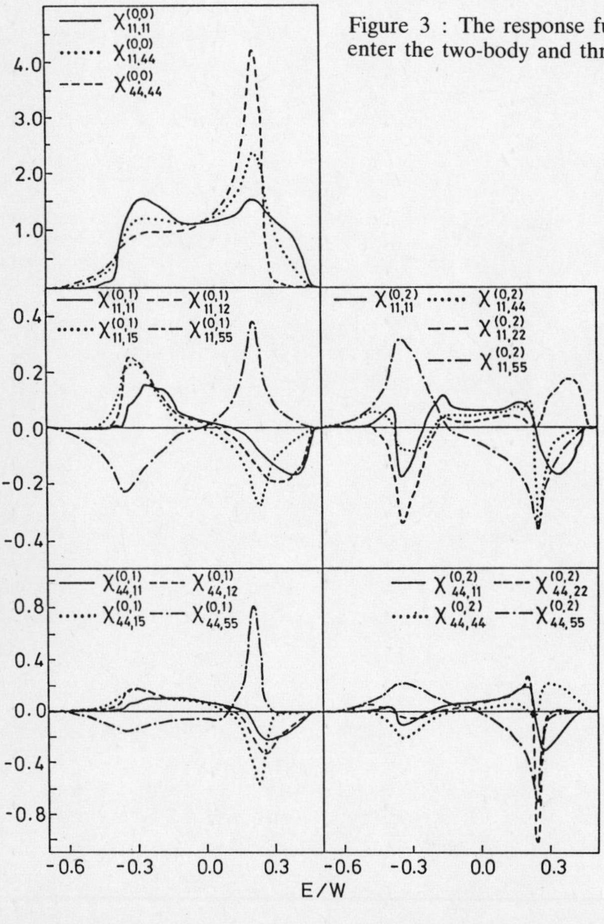

Figure 3 : The response functions involving $G^{(0)}$ which enter the two-body and three-body interactions.

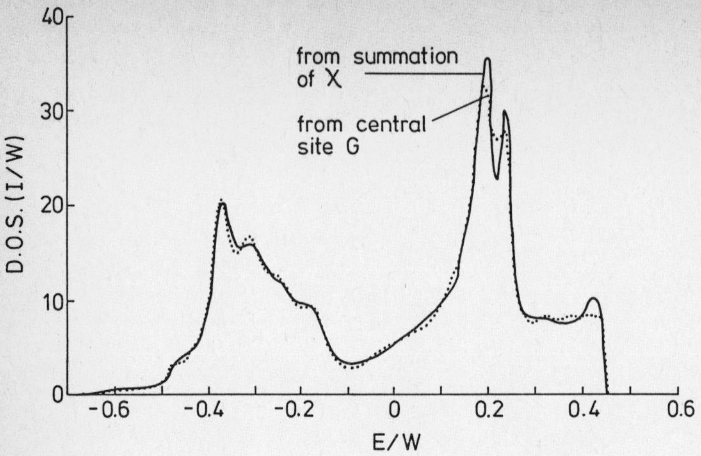

Figure 4 : The central site total density of states calculated from the site-diagonal Green function elements to 20 levels of the continued fraction, compared with the result of summing the response functions out to ten shells of neighbours.

The response functions should satisfy a sum rule[19] :

$$(-1/\pi)\ Im\ G^{pp}_{00} = \sum_{aq} \chi^{pq,qp}_{0a,a0}\,, \tag{17}$$

which provides an alternative way of calculating the local density of states. Results of evaluating (17) are shown in Figs. 4 and 5.

It appears that the density of states is fairly well reproduced by including only two shells of neighbours in (17), but longer ranged response functions reproduce more detail and are attempting to describe singularities in the density of states where convergence is poorest, notably at the E_g antibonding peak, which is split in the exact and 20 level density of states.

4 Force constants

The strong local response functions, involving $G^{(0)}G^{(0)}$, only enter the first and second neighbour force constants, shown diagramatically in Fig.6a.
The following are the explicit contributions they make in the canonical case :

$$\alpha_1 = -(200/27)(\beta^{(1)})^2(254\chi^{(0,0)}_{11,11} + 248\chi^{(0,0)}_{11,44} + 65\chi^{(0,0)}_{44,44}), \tag{18}$$

$$\beta_1 = -(200/27)(\beta^{(1)})^2(122\chi^{(0,0)}_{11,11} + 26\chi^{(0,0)}_{11,44} + 41\chi^{(0,0)}_{44,44}), \tag{19}$$

$$\alpha_2 = -100(\beta^{(2)})^2(33\chi^{(0,0)}_{11,11} + 37\chi^{(0,0)}_{44,44}), \tag{20}$$

$$\beta_2 = -100(\beta^{(2)})^2(2\chi^{(0,0)}_{11,11} + 26\chi^{(0,0)}_{11,44}). \tag{21}$$

The notation for force constants is that of [20] . The response functions involving only one on-site Green function contribute 3–body terms to the force constants, shown diagramatically in Fig.6b. Their contributions to the second neighbour force constants are as follows :

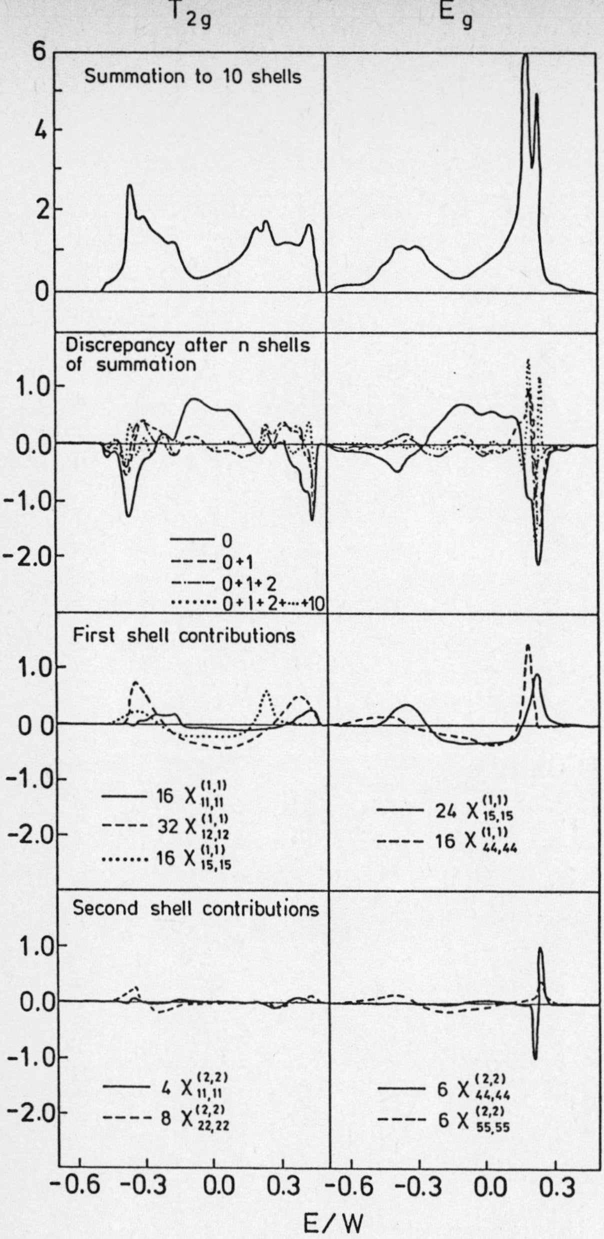

Figure 5 : Convergence of the sum rule for the local densities of states of T_{2g} and E_g symmetry out to ten shells of neighbours. The curves show the discrepancies between the summation and the on-site Green function $-(1/\pi)ImG_{pp}^{(0)}$ after cumulating the respective contributions of the central site, the first shell, the second shell and up to the tenth shell of neighbours.

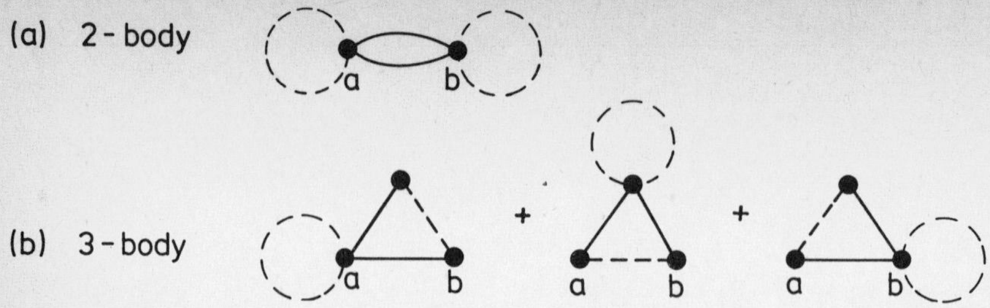

Figure 6 : Diagrams of the contributions to two-body and three-body interactions. The dashed lines indicate elements of the Green function and the solid lines indicate the perturbation ΔH.

$$\begin{aligned}\alpha_2 = &-(400/27)(\beta^{(1)})^2[-100\chi^{(0,2)}_{11,11} - 192\chi^{(0,2)}_{11,22} - 32\chi^{(0,2)}_{11,44} - 216\chi^{(0,2)}_{11,55} \\ &- 248\chi^{(0,2)}_{44,22} + 121\chi^{(0,2)}_{44,44} + 9\chi^{(0,2)}_{44,55} + (6\sqrt{3}\beta^{(2)}/\beta^{(1)})(6\chi^{(0,1)}_{11,11} + 60\chi^{(0,1)}_{11,12} \\ &+ 40\sqrt{3}\chi^{(0,1)}_{11,15} - 40\sqrt{3}\chi^{(0,1)}_{44,15} + 29\chi^{(0,1)}_{44,44})],\end{aligned} \tag{22}$$

$$\begin{aligned}\beta_2 = &-(400/27)(\beta^{(1)})^2[96\chi^{(0,2)}_{11,11} - 204\chi^{(0,2)}_{11,22} + 72\chi^{(0,2)}_{11,44} - 72\chi^{(0,2)}_{11,55} - 124\chi^{(0,2)}_{44,11} \\ &+ 124\chi^{(0,2)}_{44,22} - 37\chi^{(0,2)}_{44,44} - 93\chi^{(0,2)}_{44,55} + (6\sqrt{3}\beta^{(2)}/\beta^{(1)})(20\chi^{(0,1)}_{11,11} - 22\chi^{(0,1)}_{11,12} \\ &- 10\sqrt{3}\chi^{(0,1)}_{11,15} - 17\chi^{(0,1)}_{11,44} - 17\chi^{(0,1)}_{44,11} + 37\chi^{(0,1)}_{44,12} + \tfrac{23}{2}\sqrt{3}\chi^{(0,1)}_{44,15})].\end{aligned} \tag{23}$$

The contributions quickly become too complicated to deal with analytically, so we have evaluated the response functions and their coefficients in all the force constants by computer. For the first and second neighbours three of the main contributions in each case are shown in Fig.7 for the fully hybridised vanadium band paramaters.

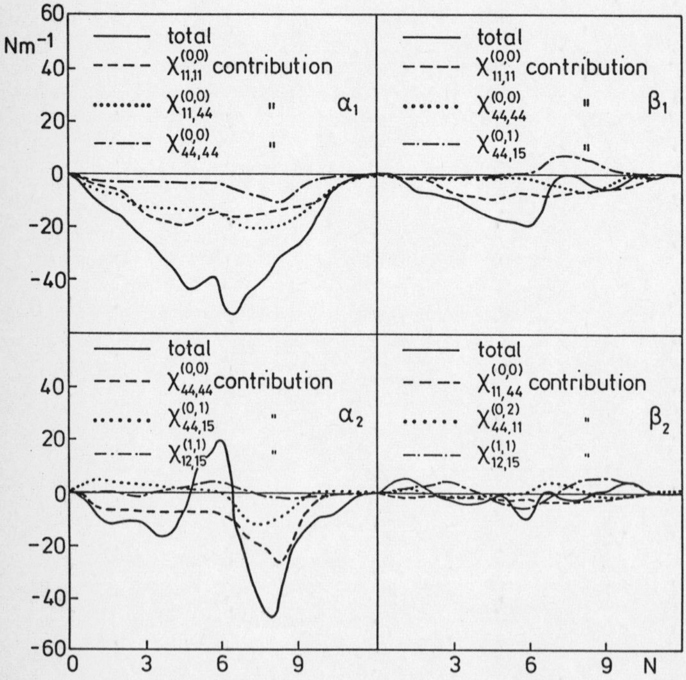

Fig. 7. Caption see opposite page

The $\chi^{(0,n)}$ terms, involving a central site Green function, are dominant in α_1 and determine the minimum around a bandfilling of 8 in α_2. However, they clearly do not dominate α_2 at a lower bandfilling, or β_2, in which more complicated three– and four-body terms are very important.

The full set of calculations with the parameters for V and the resulting phonon spectra are reproduced from [3] in Figs. (8) and (9).

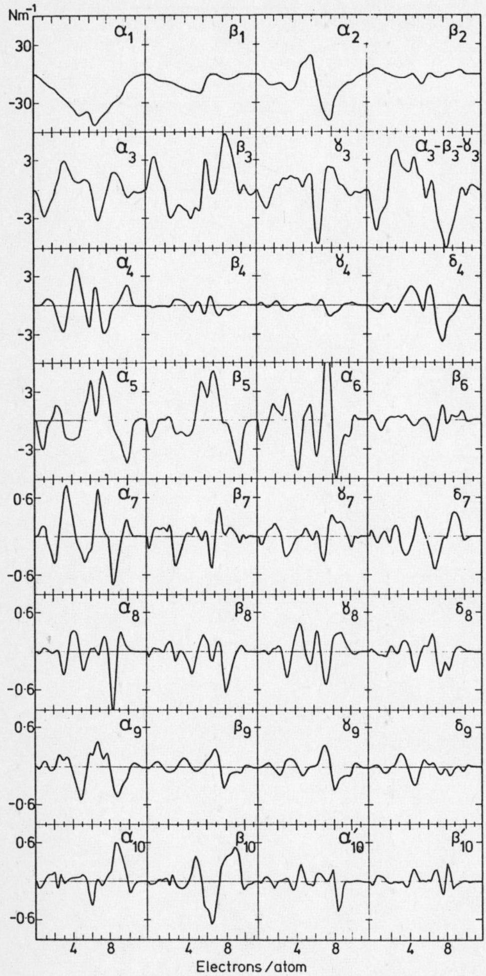

Figure 8 : The force constants for the first ten shells of neighbours as a function of band filling, omitting the empirical first and second n.n. contributions $\hat{\phi}_{\alpha\beta}$. The primed and unprimed tenth-n.n. force constants refer to the (511) and (333) neighbours respectively. Not shown : γ_{10}' and δ_{10}'.

Figure 7 : Contributions to the first and second neighbour force constants, using s, p, and d orbitals in the computation of the Green functions, as a function of band filling.

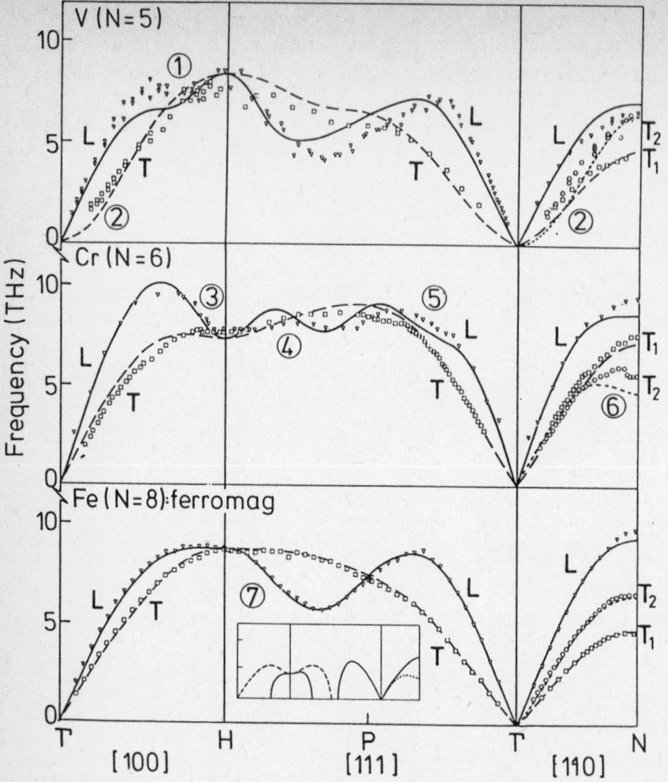

Figure 9 : Calculated phonon spectra of V, Cr and *ferromagnetic* Fe compared to experiment[21–23]. The inset shows the theoretical phonon spectrum of *nonmagnetic* Fe.

5 Discussion

The recursion method has been applied to calculate Green functions out to ten shells of neighbours in a bcc lattice, from which response functions, force constants and hence phonon spectra were calculated. The first and second neighbour force constants each contained an empirical fitting parameter; however, much of the structure in the observed phonon spectra can be attributed to the band contribution we have calculated.

It may sometimes be possible to trace the detailed structure in the phonon spectra back to particular dominant response functions. For example, at a band filling of about 7, several force constants exhibit a minimum, particularly α_2, which accounts for the softening in shear of paramagnetic iron compared to magnetic iron[24] . As Fig.7 shows, the minimum in α_2 reflects the peak in $\chi^{(0,0)}_{44,44}$ and $\chi^{(0,1)}_{44,15}$ which in turn can be attributed to the antibonding E_g peak in $(-1/\pi)\ Im\ G^{(0)}_{44}$. In other cases, no dominant dependence on one or two particular short ranged response functions is apparent. For example, the maximum in α_2 at a bandfilling of about 6 causes the maximum in the longitudinal [100] mode in Cr, as well as filling in much of the dip at $[\frac{2}{3}\ \frac{2}{3}\ \frac{2}{3}]$ in the longitudinal [111] mode compared to V. However, the rise in α_2 only develops between bandfillings of 5 and 6 as response functions involving fourth neighbours are included.

In conclusion, the recursion method enables us to use a real-space TB approach for the calculation of transition metal phonon spectra. The formalism does not depend on lattice periodicity, and so is directly applicable to surfaces, defects and impurities.

References

1. C.M.Varma and W.Weber, *Phys.Rev. B* **19** , 6142, (1979).
2. D.H.Lee and J.D.Joannopoulos, *Phys.Rev.Lett.* **48** , 1846, (1982).
3. M.W.Finnis, K.L.Kear and D.G.Pettifor, *Phys.Rev.Lett.* **52** , 291, (1984).
4. Douglas C.Allan and E.J.Mele, *Phys.Rev.Lett.* **53** , 826, (1984).
5. Y.Ohta and M.Shimizu, *J.Phys.F* **13** , 761, (1983).
6. J.C.Slater and G.F.Koster, *Phys.Rev.* **94** , 1498, (1954).
7. F.Ducastelle, *J.Phys. (Paris)* **31** , 1055, (1970).
8. G.Moraitis and F.Gautier, *J.Phys.F* **7** , 1841, (1977).
9. K.Terakura, *J.Phys.C* **11** , 469, (1978).
10. O.K.Andersen, *Phys.Rev.* **12** , 3060, (1975).
11. D.G.Pettifor, *J.Phys.F* **7** , 613, (1978).
12. R.Haydock, V.Heine and M.J.Kelly, *J.Phys.C* **8** , 2591, (1975).
13. N.Beer and D.G.Pettifor, in *Electronic Structure of Complex Systems* (Plenum, New York, to be published).
14. V.Heine and J.H.Samson, *J.Phys.F* **10** , 2609, (1980).
15. A.Bieber, F.Gautier, G.Treglia and F.Ducastelle, *Solid State Commun.* **39** , 149, (1981).
16. G.Treglia and A.Bieber, *J.Physique* **45** , 283,(1984).
17. R.Muniz, *PhD Thesis,* Dept. of Mathematics, Imperial College, (1983).
18. R.McWeeny, in *Orbital Theories of Molecules and Solids,* ed. N.H.March, (Clarendon, Oxford), p.203, (1974).
19. K.Terakura, N.Hamada, T.Oguchi and T.Asada, *J.Phys.F* **12** , 1661, (1982).
20. A.D.B.Woods, B.N.Brockhouse, R.H.March, A.T.Stewart and R.Bowers, *Phys.Rev.* **128** , 1112, (1962).
21. R.Colella and B.W.Batterman, *Phys.Rev.B* **1** , 3913, (1970).
22. W.M.Shaw and L.D.Muhlstein, *Phys.Rev.B* **4** , 969, (1971).
23. A.M.B.G.De Vallera, *PhD Thesis,* Cambridge University, (1977).
24. H.Hasegawa, M.W.Finnis and D.G.Pettifor, *J.Phys.F* (to be published).

The Recursion Method with a Non-Orthogonal Basis

R. Jones

Department of Physics, University of Exeter, Stocker Road,
Exeter EX4 4QL, United Kingdom

The problems involved in applying the recursion method to a hamiltonian with non-orthogonal basis functions are described. Methods of calculating both the local density of states and the charge density are given. The latter is most easily calculated using a block tridiagonalisation scheme. The method is illustrated for diamond.

1. Introduction

The successful approximation of exchange and correlation energies by a local density functional has created considerable interest in efficient methods of directly evaluating the charge density. The continued fraction method is known to be a numerically stable way of evaluating the diagonal Green's function of a tight-binding hamiltonian [1] and one wonders whether it can be generalised to deal with the charge density. We have explored this question [2] starting from a localised set of basis functions $\langle \underline{r}|\ell\alpha\rangle$ centred on sites α with orbital index ℓ. We suppose these functions have a finite range i.e. $\langle \underline{r}|\alpha\ell\rangle = 0$ for $\underline{r}$ sufficiently far from α.

For most calculations these basis functions are not orthogonal to each other and hence

$$S_{\ell\alpha,\ell'\alpha'} = \int d\underline{r}\ \langle \ell\alpha|r\rangle\langle r|\ell'\alpha'\rangle$$

cannot be taken to be a diagonal matrix but rather one which vanishes for α sufficiently far from α'. This raises considerable problems in evaluating $\rho(\underline{r})$. The hamiltonian matrix elements are

$$H_{\ell\alpha,\ell'\alpha'} = \int \langle \ell\alpha|\underline{r}\rangle H \langle \underline{r}|\ell'\alpha'\rangle\ d\underline{r}.$$

Suppose the λth eigenvector of H is $\langle r|\lambda\rangle$. If we expand this in the basis functions as

$$\langle r|\lambda\rangle = \sum_{\ell\alpha} C^{\lambda}_{\ell\alpha}\ \langle r|\ell\alpha\rangle \tag{1.1}$$

Then for each $\ell\alpha$

$$\sum_{\ell'\alpha'} (H_{\ell\alpha,\ell'\alpha'} - E_\lambda S_{\ell\alpha,\ell'\alpha'})\ C^{\lambda}_{\ell'\alpha'} = 0 \tag{1.2}$$

The normalisation of $\langle r|\lambda\rangle$ requires

$$\sum_{\ell\alpha} C^{\lambda *}_{\ell\alpha} S_{\ell\alpha,\ell'\alpha'} C^{\lambda}_{\ell'\alpha'} = 1 \tag{1.3}$$

We introduce the Green's function

$$G_{\ell\alpha,\ell'\alpha'}(E) = \langle \ell\alpha | 1/E^{+}S-H | \ell'\alpha' \rangle \tag{1.4}$$

which can be shown to be equal to

$$\sum_{\lambda} \frac{C^{\lambda *}_{\ell\alpha} C^{\lambda}_{\ell'\alpha'}}{(E^{+} - E_{\lambda})} . \tag{1.5}$$

The charge density is

$$\rho(\underline{r}) = 2 \sum_{\substack{\lambda\ occ. \\ \ell\alpha,\ell'\alpha'}} C^{\lambda *}_{\ell\alpha} C^{\lambda}_{\ell'\alpha'} \langle r|\ell\alpha\rangle^{*} \langle r|\ell'\alpha'\rangle \tag{1.6}$$

The factor 2 arises from spin-degeneracy. From (1.5) we see

$$\rho(\underline{r}) = -\frac{2}{\pi} \text{Imag} \int_{-\infty}^{E_F} \sum_{\substack{\ell\alpha \\ \ell'\alpha'}} \langle r|\ell\alpha\rangle^{*} G_{\ell\alpha,\ell'\alpha'} \langle r|\ell'\alpha'\rangle dE \tag{1.7}$$

Invariably we shall be interested in obtaining $\rho(\underline{r})$ in a finite region e.g. a unit cell in a crystal, or a volume surrounding a defect. In such cases only a finite number of orbitals $\ell\alpha$ contribute to the sum in (1.7). Suppose p orbitals $\ell_1\alpha_1$, $\ell_2\alpha_2$ ... $\ell_p\alpha_p$, contribute. Then we require the p(p+1)/2 elements $G_{\ell_i\alpha_i,\ \ell_j\alpha_j}$, $i \geq j$, for a problem devoid of point-group symmetry.

One could evaluate these by the usual recursion method [3] but a more efficient way was described in [2]. We shall discuss this in Section 2 below but proofs of the various relations can be found in [2]. One problem is that the method requires the repeated evaluation of $S^{-1}|k\rangle$ for various vectors $|k\rangle$. We shall describe in Section 3 the best way of accomplishing this. In Section 4 we describe an application of the method to evaluate the charge density in diamond obtaining results in good agreement with other methods.

2. Method

In the Lanczos method, vectors are generated iteratively by operating with H and ensuring that the new vector is orthogonal to all previous ones. In principle this means only the preceeding two vectors. The hamiltonian is tridiagonal in the new basis and its matrix elements are related to the coefficients that define the new vectors. Thus only two vectors need be stored at each stage. For the present problem we start with a set of p vectors $|1i)$ generated from $|\ell_i\alpha_i\rangle$, $i \leq p$, and generate new vectors $|ni)$, $n > 1$ by multiplication with $S^{-1}H$ and ensuring S-orthogonality i.e. $(mj|S|ni) = \delta_{mn}\delta_{ji}$ with all previously generated vectors. Again, in principle, this means ensuring S-orthogonality only with the current and previous sets of p vectors. In this new basis, S is a diagonal matrix and H is tridiagonal in blocks of $p \times p$ matrices i.e.

$$H_{mj,ni} = 0, \quad |m-n| > 1$$

and in addition

$$H_{ni,n+1,j} = 0, \quad j>i.$$

These matrix elements are related to the coefficients defining the new vectors. Only 2p vectors have to be stored at each stage.

Now the Green's function we seek obeys

$$G_{\ell_i\alpha_i,\ell_j\alpha_j} = \sum_{\substack{mk\\nq}} \langle\alpha_i\ell_i|mk)\, G_{mk,nq}(nq|\ell_j\alpha_j\rangle \tag{2.1}$$

Here $G_{mk,nq}$ is the resolvent of H in the new basis i.e.

$$G_{mk,nq} = (mk\left|\frac{1}{ES-H}\right|nq) \tag{2.2}$$

It appears from (2.1) that all matrix elements $G_{mk,nq}$ are required. However, a careful choice of starting vectors $|1i)$ makes this unnecessary. We choose $|1j)$ by

$$|\ell_i\alpha_i\rangle = \sum_{j\leq i} U(1i,1j)\, S|1j) \tag{2.3}$$

Here $U(1i, 1j)$ are coefficients selected so that $|1i)$ is S-orthogonal to $|1j)$, $j < i$. and $(1i|S|1i) = 1$. Substituting (2.3) into (2.1) and using the S-orthogonality of the vectors $|mk)$ gives us

$$G_{\ell_i\alpha_i,\ell_j\alpha_j} = \sum_{\substack{k\leq i\\q\leq j}} U(1i,1k)\, G_{1k,1q}\, U^*(1j, 1q) \tag{2.4}$$

Consequently, only the inverse of the first block elements need be found. These can be written in a form showing the relationship with the usual Lanczos method. Let $H_{n,m}$ denote the block matrix $(ni|H|mj)$, $i,j \leq p$. Then $G_{1k,1q}$ is the $k-q^{th}$ element of

$$1/E - H_{11} - H_{12}\,[1/E - H_{22} - H_{23}\left\{1/E - \ldots\right\}H_{32}]\,H_{21} \tag{2.5}$$

Thus the effect of blocks after the Nth is to add an energy-dependent correction to H_{NN}. Our results suggest that this fraction rapidly converges and hence enjoys the stability of the usual Lanczos method.

3. Evaluation of S^{-1}

The iterative method described above generates pN vectors by multiplication with $S^{-1}H$. It is essential to have a fast and accurate method of evaluating S^{-1}. In [2] this was accomplished by the Gauss-Seidel iterative scheme. We suppose throughout this section that the indices $\ell\alpha$ are ordered in the set i. Hence the components of a vector $|U\rangle$ and elements of S are U_i and S_{ij} respectively. Suppose we want to evaluate the vector $U\rangle$ which satisfies

$$U\rangle = S^{-1}|K\rangle \qquad (3.1)$$

Then

$$U_i = K_i - \sum_{j\leq i} (S - I)_{ij} U_j - \sum_{j>i} (S - I)_{ij} U'_j \qquad (3.2)$$

with $U'_j = U_j$.

In the Gauss-Seidel method we start by writing $U'_i = K_i$ and iterate (3.2) replacing U'_i by U_i. This is a very inefficient and time-consuming method. We now describe a much superior way based on the Cholesky decomposition of S [4]. We construct a lower triangular matrix L_{ij} by solving

$$\begin{aligned} L_{11} &= \sqrt{S_{11}} \\ S_{i1} &= L_{i1} L_{11} \qquad i>1 \\ S_{ij} &= \sum_{k\leq j} L_{ik} L_{jk} \qquad j\leq i \end{aligned} \qquad (3.3)$$

Provided the cluster size is small enough, L_{ij}, $j \leq i$, can be worked out initially and stored. U_i in equation (3.1) can then easily be found by solving firstly

$$\sum_{k\leq i} L_{ik}, q_k = K_i \quad \text{and then} \quad \sum_{k\leq i} L_{ik} U_k = q_i$$

by back substitution. This method is known to be a stable procedure [4].

4. Application to Diamond

We used a tight binding hamiltonian generated from a basis of 4 orbitals per atom [5] and a crystalline cluster of 250 atoms. The basis functions were gaussian-type orbitals each composed of 4 gaussians, and the crystalline potential (including exchange) was fitted to exponential-type functions. This allowed all matrix elements to be analytically evaluated. The basis functions were of sufficient short range that only matrix elements up to and including the third shell of atoms are important. The charge density was evaluated in a region surrounding an atom and for this purpose we chose the set of vectors $|\ell_i\alpha_i\rangle$, $1 \leq i \leq p$ corresponding to the orbitals on a central atom and its four neighbours. Then $p = 20$. We evaluated 8 blocks of the hamiltonian using the Gauss-Seidel method of evaluating S^{-1}. The matrix elements of the higher order blocks were quite regular, suggesting that the terminations discussed by Turchi, Ducastelle and Treglia [6] could be usefully employed. In fact, we diagonalised directly the 160 × 160 matrix, obtaining the eigenvalues and eigenvectors. $G_{1k,1\ell}$ could then be easily evaluated. We found that a few of the 160 eigenvalues fell outside the band edges but these had negligible amplitude on the 5 central atoms. They arise from the termination of the hamiltonian. The total charge on the central atom α_1 is

$$- \frac{2}{\pi} \text{Imag} \int^{E_F} \sum_{\substack{\ell_1 \\ \ell_j\alpha_j}} \langle \ell_1\alpha_1 | S | \ell_j\alpha_j \rangle \, G_{\ell_j\alpha_j,\ell_1\alpha_1} \, dE \tag{4.1}$$

which was within 5% of 4 and varied slightly with the number of blocks used. The integrand in (4.1) is one way of defining the local density of states [2]. The charge density along the bond between two atoms is shown in Fig. 1. The slight asymmetry between the two atoms arises because of the inequivalence between the central atom α_1 and its neighbours. The charge density agrees with one based on the usual recursion method [3] as well as an independent one [7].

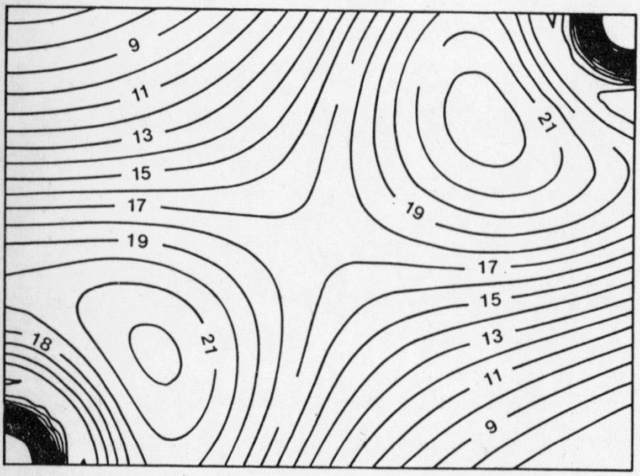

Fig. 1 Valence charge density (electrons per primitive cell) for diamond Horizontal axis is (010), and the vertical axis is (001)

Conclusions

The method introduced here generalises the Lanczos method to produce Hamiltonians of block tridiagonal form. Charge densities and local densities of states are then easily evaluated. The most time-consuming part of the calculation is the repeated evaluation, by an iterative scheme, of $S^{-1}|K\rangle$. This is bound to limit the use of this method to those problems where the off-diagonal elements of S are sufficiently small. The use of a Cholesky decomposition of S discussed in 3 would undoubtably speed up the program.

The method is particularly useful when a large cluster of atoms is needed to describe accurately the energy levels and wavefunctions of defects.

Acknowledgements

Useful discussions with M. Lewis, T. King and M. Heggie are acknowledged.

References

1. R. Haydock, V. Heine and M.J. Kelly, Solid St. Phys. 35, 216 (1980)
2. R. Jones, M.W. Lewis, Phil. Mag. B 49 95-100 (1984)
3. R. Jones and T. King, Phil. Mag. B 47 481 (1983)
4. J.H. Wilkinson: The Algebraic Eigenvalue Problem, Oxford (1965)
5. R. Jones and T. King, Phil. Mag. B 47 491 (1983)
6. P. Turchi, R. Ducastelle and G. Treglia, J. Phys. C 15 2891 (1982)
7. M.T. Yin and M.L. Cohen, Phys. Rev. B 24 6121 (1981)

Part V

Lanczos Method Applications

Hamiltonian Eigenvalues for Lattice Gauge Theories

A.C. Irving

Department of Applied Mathematics and Theoretical Physics
University of Liverpool, P.O. Box 147, Liverpool L69 3BX, United Kingdom

A brief review is presented of Lanczos sparse matrix techniques in solving Hamiltonian lattice gauge theory. The Hamiltonian approach gives direct access to many measurable quantities (such as masses) in quantum field theory. A brute force approach (initiated by Hamer and Barber) of exactly solving a spatially restricted system proves extremely powerful when coupled with scaling and renormalisation methods. The simple Lanczos (tri-) diagonalisation scheme provides a practical way of performing the exact calculation on restricted systems. The matrices are large, speed is required but accuracy cannot be sacrificed. Some successful applications are reviewed.

1. Hamiltonian lattice gauge theory objectives

1.1 Motivation

Theoretical physicists are now almost unanimous in their belief that all matter and interactions in the universe are described by quantum gauge field theories. If one can identify the fundamental theory, the remaining task is just a mixture of "chemistry" and applied mathematics. This rather extreme view undervalues the importance of extracting numerical predictions from a candidate theory in order to compare with experiment. For the electromagnetic interactions (atomic physics, solid state, chemistry, biology, etc.) this is a relatively straightforward matter due to the applicability of perturbation theory to quantum electrodynamics (QED) which has a small effective coupling ($\alpha \simeq 1/137$). For the strong interactions (nuclear physics, quark physics) this is not so because the effective coupling α_s of the relevant gauge theory quantum chromodynamics (QCD) can be very large. Lattice gauge theory provides one of the very few ways we know to extract numerical information from a quantum field theory such as QCD.

A typical gauge theory such as QED (with symmetry group U(1)) is specified by a Lagrangian from which one derives a corresponding Hamiltonian e.g.

$$H = \int d^3x\,[\underset{\sim}{E}^2 + \underset{\sim}{B}^2] \qquad (1)$$

(the total energy of the Maxwell field). A quantum theory of photons can be constructed from this by the so-called canonical quantisation procedure, in which the field variables $\underset{\sim}{E}$ and $\underset{\sim}{B}$ and the Hamiltonian become operators with some specified commutation relations determined by the group (in this case U(1)). A major goal of the particle physicist is to find the spectrum of such a (quantum) Hamiltonian operator $\hat{H}$. In the case of (1) the answer (free massless photons) is not so interesting, but for the analogous QCD Hamiltonian the spectrum is rich (protons, neutrons, pions, resonances, glueballs ....). Can one understand the experimental spectrum? Why are "free" quarks and gluons not eigenstates of $\hat{H}$?

As hinted above, the answer to such questions cannot be given by standard perturbation theory (Feynman rules). However, by restricting $\hat{H}$ to a discrete lattice in space[1], the theory becomes much better behaved and large effective couplings do not bother us. The gauge field variables $\underset{\sim}{E}$ and $\underset{\sim}{A}$ ($\underset{\sim}{B}$ = curl $\underset{\sim}{A}$) are now constrained to links (ℓ) between points in space.

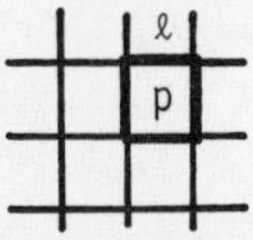

ℓ = link
p = plaquette (□)

Typically, the lattice Hamiltonian operator H now looks like

$$\hat{H} = \sum_{\ell} \hat{E}_{\ell}^{\,2} - \lambda \sum_{p} \hat{B}_{p}^{\,2} \equiv \hat{H}_o + \lambda\hat{V} \qquad (2)$$

Here $\lambda = 1/g^4$ is related to the bare coupling of the theory ($g = e_o$, the charge for QED) and the operator $\hat{B}_p^{\,2}$ typically excites the elementary plaquette (square), p, of the lattice to a higher eigenstate of $\hat{H}_o$ (sum of the link energies squared). The aim, then, is to find the low-lying eigenvalues of $\hat{H}$ by some or other technique.

The form of eqn. (2) might suggest a perturbative treatment: work in eigenstates of H_o, perturb with V and obtain a power series in λ. This is certainly a good idea when $\lambda = 1/g^4$ is small[2] (at strong coupling g), but it turns out that the interesting physics is usually near $g \simeq 0$ (weak coupling), where the effective lattice spacing or "graininess" goes away

---

1) An alternative lattice formulation in which time is also discrete is commonly used. See the contribution of M. Teper and ref. [2].

2) This is not the same as "Feynman rules" perturbation theory in the renormalised charge of the continuum theory.

(the so-called continuum limit). Ideally, one would like to know these eigenvalues at all couplings λ. One way of proceeding suggested by Hamer and Barber in 1981 [1] is something of a brute-force approach: go right ahead and diagonalise a matrix representation of $\hat{H}$! Since computers are of finite size, the suggestion is to obtain a finite matrix representation of $\hat{H}$ by considering a finite size of lattice - e.g. of linear size L = Na (where N = number of sites, a = lattice spacing). By finding the eigenvalues on a sequence of lattice sizes N one can then deduce the behaviour of an infinite size lattice (the bulk limit is $N\to\infty$). By studying the dependence of these "bulk" eigenvalues on λ (i.e. g^2) one can deduce what the answer will be in the limit $a\to 0$ (the continuum limit).

This last limit is a rather subtle one and depends on the type of theory being investigated. For a theory such as QCD which is "asymptotically free", we know that as $a\to 0$ (short distances) so must $g\to 0$ (free uncoupled particles). We also expect that at g=0 the lattice theory has a critical point where the correlation length diverges. This crucial feature makes a lattice approach to QCD worthwhile, because it means that the graininess of the lattice (set by a) becomes irrelevant at weak coupling (g small) compared to any dimensionful quantity we calculate, such as a mass whose size is controlled by the correlation length. The aim, then, is to calculate eigenvalues of $\hat{H}$ as a function of N and λ with N as large as possible and λ as near to the critical point λ_c as one can get in order to deduce the bulk limit and hence the continuum limit of the spectrum of $\hat{H}$.

One should emphasize that there are many other approaches to lattice gauge theory. A recent survey can be found in ref. [2].

1.2 Measurements

The lowest eigenvalue E_o of H tells us the vacuum energy density

$$f_o = E_o/N^d \tag{3}$$

(d is the number of spatial dimensions of the theory). From this one can extract a specific heat-like quantity

$$C = -\frac{\partial^2 f_o}{\partial\lambda^2} \tag{4}$$

Masses are just eigenvalue differences:

$$m_i = E_i - E_o \tag{5}$$

A further quantity of great interest to the lattice gauge theorist is the "string tension" T, the excess energy in the presence of two fixed charges joined by a line of "electric flux":

$$T = \frac{E_0(\text{flux string}) - E_0}{N} \tag{6}$$

Masses of different states (5) can be isolated by restricting the matrix representation of $\hat{H}$ to a particular symmetry sector of Hilbert space. Likewise E_0 (flux string) (eqn. (6)) is just the lowest eigenvalue of $\hat{H}$ restricted to basis states including a line of non-zero flux links.

In spin models (rather than gauge theories) one can also calculate things like magnetisation, susceptibilities and so on.

1.3 Physics interpretation

The phase-structure of the theory can be deduced from the mass-gap m_1 (5). Since the correlation length $\xi = 1/m_1$ (in lattice units) a vanishing m_1 signals a critical point. But on a finite lattice it cannot vanish! By using renormalisation group techniques, however, we can infer the bulk critical behaviour if it exists (e.g. calculate the Callan-Symanzik β-function from $m_1(\lambda,N)$ and find its zeros). The specific heat C also may signal a critical point. The string tension T can tell us whether a phase is "confining" or not. If the energy required to completely separate charges is infinite ($T \neq 0$) then the phase is confining. Knowing the spectrum as a function of (λ,N) and the position of any critical points, allows us to deduce the continuum physics of interest.

2. Exact eigenvalues of a finite Hamiltonian Lattice

2.1 Procedure

(A) Choose a linear size N of lattice and obtain a matrix representation of $\hat{H}_0$ and $\hat{V}$. The matrix $\langle\hat{H}_0\rangle$ is diagonal and $\langle\hat{V}\rangle$ is extremely sparse (locality) with simple entries. Take advantage of these features in storing them.

(B) Perform Lanczos (or other recursive) algorithm to tri-diagonalise and hence diagonalise $\langle\hat{H}\rangle = \langle\hat{H}_0\rangle + \lambda\langle\hat{V}\rangle$.

(C) (i) Scale eigenvalues $E_i(\lambda,N)$ to $N\to\infty$ limit by accelerated sequence algorithm [1] and/or (ii) use renormalisation procedures where one physical quantity e.g. $m_i(\lambda,N)$ is field fixed and the scale dependence is absorbed by varying the coupling (c.f. β-function in §1.3).

(D) Analyse physics.

The suggestion to use the Lanczos algorithm at step (B) was made by Roomany et al. [3] in one of the first finite lattice analyses. They tested the method on a simple Ising spin model in 1+1 dimensions (1 space + 1 time) which is of course exactly soluble by other means. In such a simple system one can achieve astonishing accuracy with minimal effort. The mass-gap $m_1(\lambda,N)$ can be calculated to, say, 12 significant figures and, using $N \le 10$ only, the bulk limits $m_1(\lambda)$ can be obtained to around 8 significant figures away from a critical point [1]. The critical coupling is correct to 5 figures and the associated critical index to around 4 or better. The largest matrix is of dimension $2^{10}/10 \simeq 100$ with 10 non-zero elements per row (of $\langle\hat{V}\rangle$) each of which is just 1! This is trivial stuff computationally.

Although this meeting has the Lanczos algorithm as its focal point, I shall say little about our implementation of it, and refer to Haydock's article [4] for general features and to our earlier work for specifics [5]. As is clear from the above outline, the actual Lanczos step (B) is a relatively minor and straightforward part of our analysis. We use no frills, just a few home-grown lines of FORTRAN describing the basic non-reorthogonalised Lanczos transformation to a chain (§§5,7 of ref. [4]).

The main advantages are (i) the economy of storage and (ii) the efficiency in finding the lowest 1 or 2 eigenvalues of $\hat{H}$ to an accuracy sufficient for our purposes (typically 10^{-10}). Even for a matrix of dimension 10^4 this usually requires no more than of order 25 Lanczos iterations. Near critical points, the accuracy in obtaining a mass-gap (5) can be poorer if convergence of E_i is slower than the rate at which orthogonalisation is lost (the introduction of ghost eigenvectors). We have not adopted the technique mentioned by Teper in his contribution, in which <u>all</u> genuine eigenvalues are accurately identified. This involves iteration of the scheme to the order of the matrix and beyond and would severely limit the practicality of our methods. In general, we are not interested in <u>all</u> the eigenvalues - only the lowest one or two.

We have used a variety of computers for our work from minis to the CRAY 1. For the latter we took care to maximise the use of vectorisation and minimise indirect addressing by using gather/scatter facilities. However, the gains over a serial machine were rather poor. We have found VAX 11/780 machines with large central memory and a very efficient virtual memory system to be highly satisfactory. On such a machine, a large computation for us would involve 9 Mbyte and last a few tens of hours in all. Examples of large scale calculations can be found in refs. [5,6,7]. We now discuss some features of such a calculation.

2.2 An example: QED in 2+1 dimensions [7]

The Hamiltonian for this model is of the form (2). Even on a finite lattice (NxN sites with periodic boundary conditions), an infinite set of basis states can be produced by the repeated action of $\hat{V}$ on the "unperturbed vacuum" (state with $E_\ell^2 = 0$ for all links ℓ). To handle this, we truncate the basis by imposing some physical cut-off (e.g. $|E_\ell| \leq L_{max}$) and calculate eigenvalues for a sequence of such cut-offs ($L_{max} = 1,2,3,\ldots$). By sequence extrapolation in L_{max} we can then estimate the "exact" finite lattice eigenvalues $E_i(\lambda,N)$. Fig. 1 shows the result of such a calculation for the energy density E_0/N^2 for $N \leq 4$. The 5x5 lattice value and the consequent bulk limit ($N \to \infty$) are indistinguishable from the 4x4 curve in this figure. Such rapid convergence of finite lattice eigenvalues is common for the ground state.

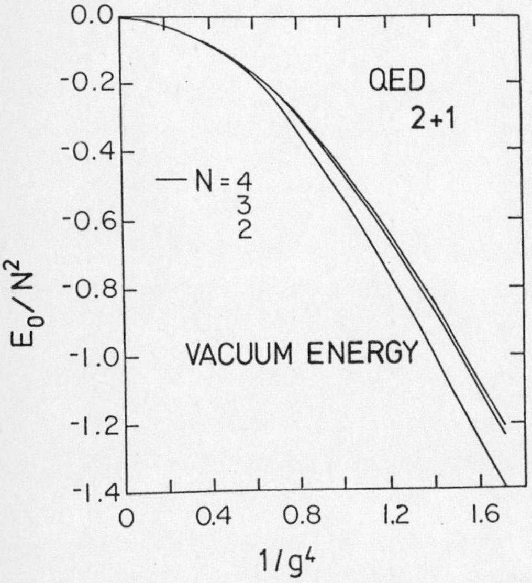

Fig. 1. The vacuum energy density for QED in D=2+1 on a sequence of finite lattices

In fig. 2 is shown the mass of the odd parity "photo-ball" of the theory. Unlike the 3+1 dimensional theory, photons can interact with each other at all values of the bare coupling g. If our world were flat, light would come in heavy lumps! There are many other interesting things one can learn about this and similar theories from a finite lattice approach [7].

2.3 Limiting factors

The main limiting factor is clearly the number of states needed to describe the theory. Even the simple Ising spin model in d+1 dimensions gives rise to 2^{N^d} states on a lattice of side N. In practice, one uses spatial

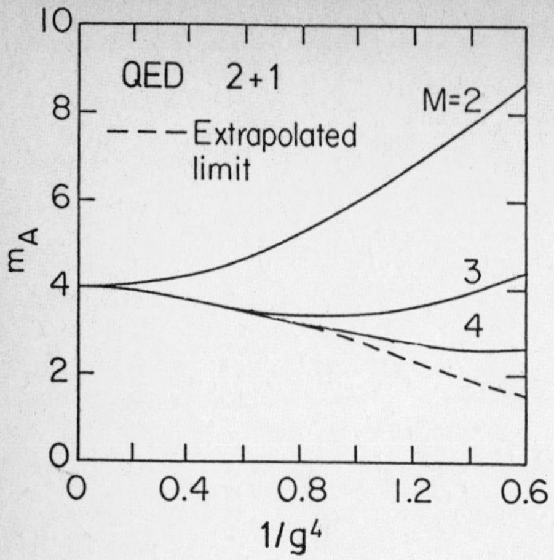

Fig. 2. The antisymmetric "photoball" mass m_A of QED in 2+1 dimensions [7]

symmetries to reduce this somewhat (e.g. for d=2 one can use N^2 translations, 4 rotations and 2 reflections). Nonetheless d ≥ 3 looks impracticable and other techniques must be employed. One such is described below.

3. Exact linked cluster expansions

In order to study more complicated theories (e.g. QCD) in higher dimensions we have developed a new algorithm [8]. Instead of considering a sequence of regular finite lattices 1^d, 2^d, 3^d, ..., we consider a sequence of linked clusters of plaquettes

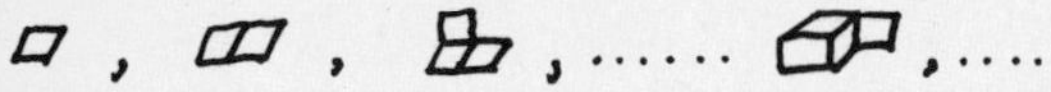 (7)

Provided we embed these correctly in the overall bulk lattice in which we are interested, we can obtain a highly convergent sequence of estimates of bulk eigenvalues [8]. The method is based on a strong coupling series linked-cluster expansion [8] but, instead, each cluster contribution is evaluated exactly using Lanczos techniques. The algorithm's acronym ELCE stands for exact linked cluster expansion.

The generation of matrix representation for $\hat{H}$ and $\hat{V}$ and their subsequent diagonalisation proceeds exactly as in steps A and B of §2.1 (except that one uses the cluster topologies of (7) rather than regular lattices). Because many topologies are involved (hundreds perhaps) one is even more dependent on the speed and efficiency of the Lanczos method. Because a considerable amount of cancellation takes place in embedding the cluster eigenvalues, the accuracy of the method is also paramount. Figure 3 shows the string tension T (eqn. 6) for QED in 2+1 and in 3+1 dimensions estimated using this sort of technique.

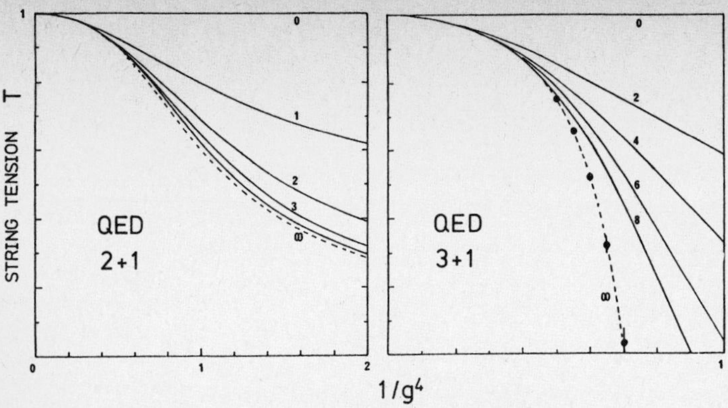

Fig. 3. The string tension (eqn. 6) for QED in 2+1 and in 3+1 dimensions [8]. The numbers are related to a cut-off on the maximum cluster size used. (∞ represents the extrapolated bulk limit)

The physics interpretation of fig. 3 is that in 2+1 dimensions, photons on a lattice self-interact and confine charges at all values of the coupling whereas in 3+1 dimension they become unconfined at some finite bare coupling ($g^2 \simeq 1.2$). At this critical point, where $T \to 0$, one can take the continuum limit as discussed in §1.1 and recover free massless continuum Maxwell electrodynamics.

Such techniques are now being used to study non-Abelian gauge theories such as QCD.

4. Concluding Remarks

The hard task in finite Hamiltonian calculations is obtaining a restricted basis and corresponding finite matrix of a manageable size. The actual matrix diagonalisation step (B) is performed in a rather straightforward way using the Lanczos algorithm. Other algorithms, such as the conjugate gradient method, are comparably efficient for the ground state but can become unreliable for higher eigenvalues [1,6].

Recently a modified Lanczos algorithm for Hamiltonian eigenvalues has been proposed by Alberty et al. [9]. They propose to diagonalise eqn. (2) by applying Lanczos successively to Hamiltonians of the form

$$\hat{H}_{n+1} = \hat{H}_n + (\lambda/M)\hat{V} \quad \text{where}$$
$$H_1 = \hat{H}_0 + (\lambda/M)\hat{V} \quad \text{and}$$
$$\hat{H} = \hat{H}_M.$$

For fixed λ, the "perturbing" piece of the Hamiltonian can be made small (λ/M). Because of this, they can obtain an nth estimate of an eigenvalue quite accurately by restricting the basis used in the diagonalisation of $\hat{H}_n$ to those states connected to the $(n-1)^{th}$ ground state by a single action of $(\lambda/M)\hat{V}$. So, at each step n, a sequence of Lanczos substeps is used to obtain the nth eigenvalues and eigenvector. By repeating this procedure ($n = 1,2,\ldots,M$) a diagonalisation of $H(\equiv H_M)$ itself is finally achieved. The authors claim an improved convergence-rate as compared with the standard Lanczos procedure and point out an added attraction that much smaller matrices need be diagonalised. One must still, at some stage, manipulate vectors corresponding to the full Hilbert space ($\hat{H}_M$) but since the matrix multiplication only involves subspaces, the problems of large storage and indirect addressing are presumably lessened. Tests of a 1+1 dimensional model have been presented [9]. It will be interesting to see if the method has anything to offer the higher dimensional theories discussed above.

I have no doubt that the Lanczos recursion method will continue to pop up in lattice gauge theory calculations for some time to come. Its simplicity and reliability, as much as anything else, will ensure that.

1. C.J. Hamer and M.N. Barber, J. Phys. A14, 241, 259, 2009 (1981)
2. J.B. Kogut, Rev. Mod. Phys. 55, 775 (1983)
3. H. Roomany, H.W. Wyld and L.E. Holloway, Phys. Rev. D21, 1557 (1980)
4. R. Haydock, Solid State Physics 35, 215 (1980)
5. A.C. Irving and A. Thomas, Nucl. Phys. B200 [FS4], 424 (1982); B215 [FS7], 23 (1983)
6. C.J. Hamer, J. Phys. A16, 1257 (1983)
7. A.C. Irving, J.F. Owens and C.J. Hamer, Phys. Rev. D28, 2059 (1983)
8. A.C. Irving and C.J. Hamer, Nucl. Phys. B230 [FS10], 361 (1984); B235 [FS11], 358 (1984)
9. J. Alberty, J. Greensite and A. Patkos, Phys. Lett. 138B, 405 (1984)

The Lanczos Method in Lattice Gauge Theories

I.M. Barbour, N.-E. Behilil, and P.E. Gibbs
Department of Natural Philosophy, University of Glasgow
Glasgow, United Kingdom
G. Schierholz
Deutsches Elektronen-Synchrotron DESY, D-2000 Hamburg, Fed. Rep. of Germany
M. Teper
L.A.P.P., Annecy

We present a modified version of the Lanczos algorithm as a computational method for tridiagonalising large sparse matrices, which avoids the requirement for large amounts of storage space. It can be applied as a first step in calculating eigenvalues and eigenvectors or for obtaining the inverse of a matrix row by row. Here we describe the method and apply it to various problems in lattice gauge theories. We have found it to have excellent convergence properties. In particular, it enables us to do lattice calculations at small and even zero quark mass.

1. Introduction

The lattice is a four-dimensional cubic array of $n_x^3 n_t$ sites, which represents a discrete approximation to a volume of space-time. The gauge fields of QCD are represented by a configuration of 3×3 special unitary matrices [SU(3)], which are located on the links of the lattice joining neighbouring sites. The fermion field for the quarks is a colour triplet of anticommuting variables, which we place on the lattice sites.

We are supplied with configurations of the gauge fields generated by Monte Carlo methods. The dynamics of the quarks in these background fields is derived from the action

$$S_F = -\bar{\psi}(M + 2ma)\psi \quad , \tag{1}$$

where ψ is the fermion field, m is the quark mass, a is the physical distance between lattice sites and M is the fermion matrix — an operator depending on the gauge configuration, which can be represented by an $N\times N$ sparse matrix ($N = 3n_x^3 n_t$). There are various possible forms of M that can be used. We use Kogut-Susskind fermions, so that M takes the following form:

$$\bar{\psi}M\psi = \sum_{n,\mu} \bar{\psi}_n U_{n,n+\hat{\mu}} f_{n,n+\hat{\mu}} \psi_{n+\hat{\mu}} - \text{hermitian conjugate} \quad , \tag{2}$$

where ψ_n is the single component colour triplet sited a $n = (x_1, x_2, x_3, t)$,

μ indexes the 4 directions in space-time, $\hat{\mu}$ is a displacement vector of length a in direction μ, $U_{n,n+\hat{\mu}}$ is the 3×3 gauge matrix joining sites n to $n+\hat{\mu}$ and the sign factors are

$$f_{n,n+\hat{\mu}} = \begin{cases} 1 & , \quad \mu = x_1 \\ (-1)^{x_1} & , \quad \mu = x_2 \\ (-1)^{x_1+x_2} & , \quad \mu = x_3 \\ (-1)^{x_1+x_2+x_3} & , \quad \mu = t \quad . \end{cases} \tag{3}$$

For the gauge field action we assume periodic boundary conditions, while for the fermionic action we shall use antiperiodic boundary conditions. This is accomplished by including an extra factor -1 in $f_{n,n+\hat{\mu}}$ between the first and last μ-plane.

Calculation of physical quantities, such as the chiral condensate and the hadron spectrum, require the calculation of the inverse of $M+2ma$. Various methods have been used in the past to do this. Gaussian elimination cannot be used because of the problem of "fill in", i.e., during the calculation the zero elements in the sparse matrix $M+2ma$ are filled in, so that vast amounts of storage and computation are required. This can be kept to a minimum by choosing suitable pivots but could not be used on lattices any bigger than $4^3 \cdot 8$. Hopping parameter expansions, Gauss-Seidel iteration and pseudo-boson methods have been tried with a certain amount of success. The drawback of these methods is that convergence is very slow at small quark masses. The conjugate gradient method for inverting matrices has been applied successfully [1,2] but again its convergence at low mass could be improved. The Lanczos method, as we present it here, has better convergence at small mass. It also has the advantage that, for one run of the Lanczos algorithm, calculations can be done at many different masses in contrast to the conjugate gradient algorithm, where each different mass requires a separate calculation.

2. The Hermitian Lanczos Algorithm

First we will describe how to tridiagonalise a hermitian matrix H (e.g., $H = iM$) of size $N \times N$. We seek a unitary transformation:

$$X^+HX = T \quad , \quad X^+X = 1 \quad , \tag{4}$$

where T is tridiagonal, real and symmetric:

$$T = \begin{bmatrix} \alpha_1 & \beta_1 & & & \\ \beta_1 & \alpha_2 & \beta_2 & & \\ & \beta_2 & \alpha_3 & & \\ & & \ddots & \ddots & \beta_{N-1} \\ & & & \beta_{N-1} & \alpha_N \end{bmatrix} \quad . \tag{5}$$

We write X as a series of column vectors:

$$X = (x_1, x_2, \ldots, x_N) \quad .$$

These are the Lanczos vectors and they are orthonormal:

$$x_i^+ x_j = \delta_{ij} \quad , \qquad \text{and} \tag{6}$$

$$HX = XT \Leftrightarrow \begin{array}{l} Hx_1 = \alpha_1 x_1 + \beta_1 x_2 \quad , \\ Hx_i = \beta_{i-1} x_{i-1} + \alpha_i x_i + \beta_i x_{i+1} \quad , \qquad 2 \leqslant i \leqslant N-1 \quad , \\ Hx_N = \beta_{N-1} x_{N-1} + \alpha_N x_N \quad . \end{array} \tag{7}$$

These are the Lanczos equations, which can be used recursively to calculate all the α_i, β_i and x_i. Chose x_1 to be any unit vector. Take the scalar product of x_1 with the first Lanczos equation and use the orthonormality of the Lanczos vectors to obtain α_1:

$$\alpha_1 = x_1^+ H x_1 \quad . \tag{8}$$

α_1 is assured to be real because H is hermitian. Next calculate

$$\beta_1 x_2 = Hx_1 - \alpha_1 x_1 \tag{9}$$

and use $x_2^+ x_2 = 1$ to obtain β_1 and x_2 (we can take either sign for β_1). Continue in a similar way with all the other equations in turn:

$$\alpha_i = x_i^+ H x_i$$

$$\beta_i x_{i+1} = Hx_i - \beta_{i-1} x_{i-1} - \alpha_i x_i \quad . \tag{10}$$

This defines the Lanczos algorithm for calculating all α_i, β_i and x_i. We can easily check that the hermitian nature of H ensures that all the Lanczos vectors are then orthogonal if the calculation can be done without rounding errors. For example,

$$\begin{aligned} x_1^+ x_2 &= \frac{1}{\beta_1} x_1^+ (Hx_1 - \alpha_1 x_1) \\ &= \frac{1}{\beta_1} (\alpha_1 - \alpha_1) = 0 \quad . \end{aligned} \tag{11}$$

Furthermore, once we have calculated

$$\alpha_N = x_N^+ H x_N \quad , \tag{12}$$

the calculation is then complete, and the last equation is automatically satisfied because we can show that

$$u = Hx_N - \beta_{N-1}x_{N-1} - \alpha_N x_N \tag{13}$$

is orthogonal to all the Lanczos vectors and must therefore be zero. In fact, a good check on the accuracy of the calculations is that

$$\beta_N = |u| = 0 \quad . \tag{14}$$

Apart from rounding errors there is only one thing that can cause the algorithm to fail. That is, if some β is zero, then we will have a division by zero. This will happen if the first Lanczos vector x_1 was chosen to be orthogonal to some eigenvector of H, and it is inevitable if H has a degenerate eigenvalue. The solution is to choose the next x_i to be any unit vector orthogonal to all the previous ones and continue the calculation. This may be difficult to implement in practice, but since we have never encountered this situation we have ignored it.

The advantage of the Lanczos algorithm over other methods is that it does not require the matrix H to be stored in a large $N \times N$ array which is "filled in" by the calculation, even if H has a large number of zero elements. We only require storage space for about 3 Lanczos vectors and a subroutine to multiply a vector by H. If H is sparse, or even a product of a few sparse matrices, then the multiplication can be done quickly and with a minimum of storage space. All but the last two Lanczos vectors can be thrown away after each iteration. If they are ever needed again (e.g., if our aim is to calculate eigenvectors of H), then they can be recalculated as they are needed, without resorting to large storage. In some cases it may be more effective to write them onto disk.

Befor we can use the Lanczos algorithm, we must overcome the problem of rounding errors. If it is applied to large matrices, we find that $\beta_N \neq 0$ due to rounding errors. This is due to loss of orthogonality between the first few Lanczos vectors and the last ones. These errors tend to build up exponentially, so that no matter what precision is used in the calculation we soon find that after each iteration the last x_i is not orthogonal to x_1. The most straightforward way to overcome this is to reorthogonalise. When we have calculated a new Lanczos vector x_i, it can be made orthogonal to an earlier vector x_j by a projection

$$x_i \rightarrow x_i - x_j(x_j^+ x_i) \quad . \tag{15}$$

x_i can be reorthogonalised against each previous vector in turn. Then, provided we have not lost too much orthogonality, the rounding errors will be reduced. Usually this does not need to be done after every iteration, unless there are many very close eigenvalues. Unfortunately, reorthogonalisation greatly slows down the calculation and means that all the Lanczos vectors must be stored. Usually it is necessary to use an external sequential disk file for this, and it is impractical to reorthogonalise for $N \gtrsim 1000$.

Fortunately, it is possible to use the Lanczos method without reorthogonalisation, and this enables us to deal with much larger matrices. We allow the Lanczos algorithm to proceed beyond the Nth iteration, calculating new Lanczos vectors and α's and β's until we have done $\tilde{N}$ iterations. Then

$$\begin{aligned} Hx_1 &= \alpha_1 x_1 + \beta_1 x_2 \quad , \\ Hx_i &= \beta_{i-1} x_{i-1} + \alpha_i x_i + \beta_i x_{i+1} \quad , \quad 2 \leqslant i \leqslant \tilde{N} \quad . \end{aligned} \tag{16}$$

The α's and β's now form an $\tilde{N}\times\tilde{N}$ tridiagonal form $\tilde{T}$ with $\tilde{N}$ eigenvalues $\tilde{\lambda}_r$, from which we can sort out the eigenvalues λ_i of H. Write the Lanczos equations as

$$H\tilde{X} = \tilde{X}\tilde{T} + R \quad . \tag{17}$$

$\tilde{X}$ is the $N \times \tilde{N}$ matrix formed with the Lanczos vectors as its columns, and R is the remainder due to rounding errors and stopping at $\tilde{N}$ iterations. The first $\tilde{N}$-1 columns of R will be small errors, but the last will be approximately $\beta_i x_{i+1}$. Suppose an eigenvalue $\tilde{\lambda}_r$ of $\tilde{T}$ has eigenvector $\tilde{e}_r$:

$$\tilde{T}\tilde{e}_r = \tilde{\lambda}_r \tilde{e}_r \quad , \tag{18}$$

$$H\tilde{X}\tilde{e}_r = \tilde{X}\tilde{T}\tilde{e}_r + R\tilde{e}_r \cong \tilde{\lambda}_r \tilde{X}\tilde{e}_r + \beta_{\tilde{N}} x_{\tilde{N}+1} \tilde{e}_{ri} \quad . \tag{19}$$

Therefore $\tilde{\lambda}_r$ will be an eigenvalue of H with eigenvector $\tilde{X}\tilde{e}_r$ provided the last component of $\tilde{e}_r$ is very small. We have found empirically that if $\tilde{N}$ is sufficiently large, then all eigenvalues of H will converge as eigenvalues of $\tilde{T}$. But $\tilde{T}$ will also have spurious eigenvalues, which are not eigenvalues of H because $\tilde{e}_{ri}$ is large. Some eigenvalues of H may converge very fast and can be obtained from $\tilde{T}$ when $\tilde{N}$ is still much smaller than N. By the time $\tilde{N}$ is large enough for all the eigenvalues to have converged, the faster ones will appear many times as eigenvalues of $\tilde{T}$. These ghosts can be recognized because we assume H to be nondegenerate. The spurious eigenvalues of $\tilde{T}$ can also be recognized by comparing with the eigenvalue of the tridiagonal matrix $\hat{T}$ formed from the first $\tilde{N}$-1 iterations. The real eigenvalues of H will be eigenvalues of $\hat{T}$ as well as $\tilde{T}$, but $\hat{T}$ will have different spurious eigenvalues. This is because the last component of their eigenvectors is large and is therefore greatly affected by removing the last α and β.

The eigenvalues of $\tilde{T}$ can be found by the standard method for Sturm sequences. This is an algorithm which quickly tells us how many eigenvalues $\tilde{T}$ has less than a given value λ. It can be used to find the nth eigenvalue of $\tilde{T}$ by a series of bisections, starting from an interval known to contain all the eigenvalues. Sturm sequences can also be used to quickly determine which are the spurious eigenvalues. Once we have an eigenvalue $\tilde{\lambda}_r$ of $\tilde{T}$, we can find how many eigenvalues of $\hat{T}$ there are in a neighbourhood $[\lambda_r - \varepsilon, \lambda_r + \varepsilon]$ of λ_r. If there are none, then λ_r is spurious.

We have not made any general studies of how the eigenvalues converge for different matrices, but we note the following two points on convergence:

(i) In any region where the eigenvalues are relatively well separated, they will converge fastest, and may appear with many ghosts before all other eigenvalues have converged. In some applications only these eigenvalues are needed, and it is only necessary to do a small number of iterations to obtain them.

(ii) If some eigenvalues are relatively small or close together, then a correspondingly high precision is needed for the α's, β's and Lanczos vectors, otherwise they will not all converge no matter how many iterations are done.

For further details see [3].

3. Application to $\langle\bar{\psi}\psi\rangle$ Calculations in Lattice QCD

We have applied the hermitian Lanczos algorithm to the study of chiral symmetry breaking in lattice QCD. For a given configuration of gauge links on a lattice of N_L sites we need to calculate the trace of the inverse of an operator $M + 2ma$, where m is the quark mass. M is antihermitian and acts on a fermion field with 3 colour components at each lattice site.

$$\langle\bar{\psi}\psi\rangle = \frac{1}{N_L} \mathrm{Tr}\{M + 2ma\}^{-1} \tag{20}$$

We are interested in the behaviour of $\langle\bar{\psi}\psi\rangle$ as ma becomes small. M can be represented by a large square matrix of size $N = 3N_L$ squared. But since M only connects field components which are joined by a lattice link, each row or column of M has only 24 components which are nonzero. Therefore it is very sparse, and the Lanczos algorithm can be used to obtain all eigenvalues of iM. Then

$$\langle\bar{\psi}\psi\rangle = \frac{3}{N} \sum_k \frac{1}{-i\lambda_k + 2ma} \quad . \tag{21}$$

The lattice has an even number of sites in each direction so that M has the following block structure between odd and even sites:

$$iM = \begin{pmatrix} 0 & \hat{M} \\ \hat{M}^{+} & 0 \end{pmatrix} . \tag{22}$$

This implies that the eigenvalues of M come in plus and minus pairs because

$$\begin{vmatrix} \lambda & \hat{M} \\ \hat{M}^{+} & \lambda \end{vmatrix} = |\lambda^2 - \hat{M}\hat{M}^{+}| = 0 \quad ,$$

$$\Rightarrow \quad \langle\bar{\psi}\psi\rangle = \frac{3}{N}\frac{1}{2}\sum_k \left(\frac{1}{-i\lambda_k + 2ma} + \frac{1}{i\lambda_k + 2ma}\right)$$
$$= \frac{3}{N}\sum_k \frac{2ma}{\lambda_k^2 + (2ma)^2} \quad . \tag{23}$$

For large lattices,

$$\langle\bar{\psi}\psi\rangle \cong 3\int_{-\infty}^{+\infty} d\lambda \, \frac{2ma\rho(\lambda)}{\lambda^2 + (2ma)^2} \quad , \tag{24}$$

where $\rho(\lambda)$ is the normalized spectral density. As $ma \to 0$ this gives

$$\langle\bar{\psi}\psi\rangle \cong 3\pi\rho(0) \quad . \tag{25}$$

Therefore we only need the eigenvalues close to zero.

The odd-even block structure of iM leads to a simplification in the Lanczos algorithm if we choose the initial Lanczos vector to be nonzero only on even sites:

$$x_1 = \begin{pmatrix} \hat{x}_1 \\ 0 \end{pmatrix} . \tag{26}$$

Then

$$\alpha_1 = x_1^{+} iMx_1 = (\hat{x}_1^{+},0)\begin{pmatrix} 0 & \hat{M} \\ \hat{M}^{+} & 0 \end{pmatrix}\begin{pmatrix} \hat{x}_1 \\ 0 \end{pmatrix} = 0 \quad ,$$

$$\beta_1 x_2 = iMx_1 - \alpha_1 x_1 = iMx_1 = \begin{pmatrix} 0 \\ \hat{M}^{+}x_1 \end{pmatrix} . \tag{27}$$

So x_2 will be nonzero only on odd sites. Continuing in a similar fashion we find that all the α's are zero, and the Lanczos vectors are all half zero:

$$x_{2k-1} = \begin{pmatrix} \hat{x}_{2k-1} \\ 0 \end{pmatrix}, \quad x_{2k} = \begin{pmatrix} 0 \\ \hat{x}_{2k} \end{pmatrix}, \quad k = 1, \ldots, \frac{1}{2}\tilde{N} \quad . \tag{28}$$

The Lanczos equations take the form

$$\hat{M}^{+}\hat{x}_1 = \beta_1\hat{x}_2 \quad ,$$
$$\hat{M}\hat{x}_{2k} = \beta_{2k-1}\hat{x}_{2k-1} + \beta_{2k}\hat{x}_{2k+1} \quad , \tag{29}$$
$$\hat{M}^{+}\hat{x}_{2k+1} = \beta_{2k}\hat{x}_{2k} + \beta_{2k+2}\hat{x}_{2k+2} \quad .$$

We have applied this algorithm to the study of chiral symmetry breaking on lattice configurations of size 8^4 [4] and to the study of chiral symmetry restoration at finite temperature on lattice configurations of sizes $12^3 \cdot 4$ and $12^3 \cdot 6$ [5]. We used $\tilde{N} = 2N$ to obtain all the eigenvalues. In the case of the $12^3 \cdot 4$ lattice there were 20756 of them. From this we were able to establish that the small eigenvalues were sufficient for our purposes. These could be obtained after $\tilde{N} = 1/4\ N$ iterations. In the Lanczos algorithm we used double precision (8 bytes) arithmetic, even though the elements of the matrix were only supplied to single precision, but for the Sturm sequences we were able to reduce this to 6 byte arithmetic. Eigenvalues of the tridiagonal form which differed by less than 1 part in 10^8 were taken to be ghosts, and those that moved by more than 1 part in 10^6 when the last α and β were removed were taken to be spurious. This left precisely the correct number of eigenvalues when $\tilde{N} = 2N$, and the sum of their squares checked with the original matrix to 8 significiant figures.

4. The Nonhermitian Lanczos Method

The Lanczos method can be generalized to include nonhermitian matrices. We require a similarity transform,

$$X^{-1} H X = T \quad . \tag{30}$$

T will not in general be hermitian, since H may have complex eigenvalues, but we can make it symmetric:

$$T = \begin{pmatrix} \alpha_1 \beta_1 & & \\ \beta_1 \alpha_2 \beta_2 & & \\ & \beta_2 \ddots & \\ & & \ddots \end{pmatrix} \quad , \tag{31}$$

with the α's and β's complex. We have the same Lanczos equations as before:

$$Hx_1 = \alpha_1 x_1 + \beta_1 x_2 \quad ,$$

$$Hx_i = \beta_{i-1} x_{i-1} + \alpha_i x_i + \beta_i x_{i+1} \quad ,$$

but X may be nonunitary, so we must also generate its inverse:

$$Y = X^{-1+} \quad , \quad H^+ Y = Y T^+ \quad , \quad Y^+ X = 1 \quad . \tag{32}$$

The columns of Y can be calculated with the additional Lanczos equations

$$\begin{aligned} H^+ y_1 &= \alpha_1^* y_1 + \beta_1^* y_2 \quad , \quad Y = (y_1, \ldots, y_N) \quad , \\ H^+ y_i &= \beta_{i-1}^* y_{i-1} + \alpha_i^* y_i + \beta_i^* y_{i+1} \quad , \end{aligned} \tag{33}$$

$$y_i^{\;+} x_i = \delta_{ij} \quad . \tag{34}$$

The s are calculated from

$$\alpha_i = y_i^+ H x_i \quad . \tag{35}$$

The β's and the Lanczos vectors x_{i+1}, y_{i+1} come from

$$\begin{aligned} \beta_i x_{i+1} &= H x_i - \alpha_i x_i - \beta_{i-1} x_{i-1} \quad , \\ \beta_i^* y_{i+1} &= H^+ y_i - \alpha_i^* y_i - \beta_{i-1}^* y_{i-1} \quad , \end{aligned} \tag{36}$$

$$\beta_i^2 = (\beta_i^* y_{i+1})^+ (\beta_i x_{i+1}) \quad . \tag{37}$$

We need to take a complex square root to calculate β_i.

As with the hermitian case, the orthogonality between the Lanczos vectors is rapidly lost due to rounding errors. Again we can continue the algorithm past N iterations and hope that the eigenvalues will still converge. Unfortunately, there is no easy way of generalising the method of Sturm sequences to a complex symmetric tridiagonal form [6]. The only alternative is to construct the coefficients of the characteristic polynomial and find its roots, but this does not work well for $N \gtrsim 100$. Therefore we are forced to use reorthogonalisation. Each x_i calculated must be reorthogonalised against all previous y_i, and similarly each y_i against the previous x_i. Once we have the tridiagonal form, the trace of the inverse can be calculated relatively quickly and can be repeated with different constants added to the diagonal.

5. Matrix Inversion

The Lanczos algorithm can be applied in a different way to invert a matrix column by column. This applies to both hermitian and nonhermitian matrices, though we have only tried it out for hermitian matrices. Generate the Lanczos equations as before without reorthogonalisation:

$$H x_1 = \alpha_1 x_1 + \beta_1 x_2 \quad , \quad H x_i = \beta_{i-1} x_{i-1} + \alpha_i x_i + \beta_i x_{i+1} \quad . \tag{38}$$

We shall aim to calculate $H^{-1} x_1$ as an infinite series in the Lanczos vectors,

$$H^{-1} x_1 = c_1 x_1 + c_2 x_2 + \ldots \quad , \tag{39}$$

by using the Lanczos equations iteratively. After k iterations of the Lanczos algorithm we will have constructed k terms in the series with a remainder term involving $H^{-1} x_k$ and $H^{-1} x_{k+1}$:

$$H^{-1}x_1 = v_k + a_k H^{-1}x_k + b_k H^{-1}x_{k+1} \quad ,$$

$$v_k = \sum_{i=1}^{k} c_i x_i \quad . \tag{40}$$

The next Lanczos equation can be used to eliminate $H^{-1}x_k$:

$$H^{-1}x_k = \frac{1}{\beta_k} x_{k+1} - \frac{\alpha_{k+1}}{\beta_k} H^{-1}x_{k+1} - \frac{\beta_{k+1}}{\beta_k} H^{-1}x_{k+2} \tag{41}$$

$$\Rightarrow H^{-1}x_1 = v_k + \frac{a_k}{\beta_k} x_{k+1} + \left(b_k - \frac{\alpha_{k+1}}{\beta_k} a_k\right) H^{-1}x_{k+1} - \frac{\beta_{k+1}}{\beta_k} a_k H^{-1}x_{k+2} \quad , \tag{42}$$

giving the following recurrence relations:

$$v_{k+1} = v_k + \frac{a_k}{\beta_k} x_{k+1} \quad ,$$

$$\begin{bmatrix} a_{k+1} \\ b_{k+1} \end{bmatrix} = \begin{bmatrix} -\dfrac{\alpha_{k+1}}{\beta_k} & 1 \\ -\dfrac{\beta_{k+1}}{\beta_k} & 0 \end{bmatrix} \begin{bmatrix} a_k \\ b_k \end{bmatrix} \quad . \tag{43}$$

Initival values for v_1, a_1 and b_1 can be obtained from the first equation if $\alpha_1 \neq 0$:

$$H^{-1}x_1 = \frac{1}{\alpha_1} x_1 - \frac{\beta_1}{\alpha_1} H^{-1}x_2 \quad , \tag{44}$$

but we can also start from the identity

$$H^{-1}x_1 = H^{-1}x_1 \quad . \tag{45}$$

As we shall see, this ambiguity in initial conditions is important for convergence of the series. Combining (44) and (45) gives the most general starting conditions:

$$\left(r - \frac{\alpha_1}{\beta_1} s\right) H^{-1}x_1 = -\frac{s}{\beta_1} x_1 + rH^{-1}x_1 + sH^{-1}x_2 \tag{46}$$

$$\Rightarrow v_1 = -\frac{s}{r - \dfrac{\alpha_1}{\beta_1} s} \frac{x_1}{\beta_1} \quad , \tag{47}$$

$$\begin{pmatrix} a_1 \\ b_1 \end{pmatrix} = \frac{1}{r - \dfrac{\alpha_1}{\beta_1} s} \begin{pmatrix} r \\ s \end{pmatrix} \quad . \tag{48}$$

The parameters r and s must be left undetermined until the end of the calculation, when they can be chosen in such a way that the remainder term is small:

$$\begin{pmatrix} a_k \\ b_k \end{pmatrix} = \frac{1}{r - \frac{\alpha_1}{\beta_1} s} \, \Pi_k \begin{pmatrix} r \\ s \end{pmatrix} \quad , \tag{49}$$

where the 2×2 matrix Π_k can be calculated from the recurrence relations

$$\Pi_1 = \begin{pmatrix} 1 & 0 \\ 0 & 1 \end{pmatrix} \quad , \tag{50}$$

$$\Pi_{k+1} = \begin{bmatrix} -\dfrac{\alpha_{k+1}}{\beta_k} & 1 \\ -\dfrac{\beta_{k+1}}{\beta_k} & 0 \end{bmatrix} \Pi_k \quad . \tag{51}$$

For convergence of the series we require

$$\begin{pmatrix} a_k \\ b_k \end{pmatrix} \to \begin{pmatrix} 0 \\ 0 \end{pmatrix} \quad . \tag{52}$$

This would happen for any choice of initial conditions if

$$\Pi_k \to 0 \quad , \tag{53}$$

but

$$\det \Pi_k = \prod_{i=1}^{k-1} \det \begin{bmatrix} -\dfrac{\alpha_{i+1}}{\beta_i} & 1 \\ -\dfrac{\beta_{i+1}}{\beta_i} & 0 \end{bmatrix} = \prod_{i=1}^{k-1} \frac{\beta_{i+1}}{\beta_i} = \frac{\beta_k}{\beta_1} \quad . \tag{54}$$

Therefore, unless we have a β equal to zero, we cannot have $\Pi_k \to 0$. However, if one eigenvalue of Π_k tends to zero, we can take $\begin{pmatrix} r \\ s \end{pmatrix}$ to be the corresponding eigenvector, and this will suffice to make the remainder term small. Since we will not know r and s until the end, we must compute v_k as a linear combination:

$$v_k = \frac{1}{r - \frac{\alpha_1}{\beta_1} s} \, \sigma_k \begin{pmatrix} r \\ s \end{pmatrix} \quad . \tag{55}$$

σ_k has two component vectors and is generated from the following relations:

$$\sigma_1 = \left(0 \; , \; -\frac{x_1}{\beta_1}\right) \quad , \tag{56}$$

$$\sigma_{k+1} = \sigma_k + \left(\frac{x_{k+1}}{\beta_k}, 0\right)\Pi_k \quad . \tag{57}$$

If we now proceed to calculate Π_k and σ_k from (50,51,56,57) naively, then we would run into a problem with rounding errors. As one eigenvalue of Π_k converges to zero, the other tends to diverge, because the determinant fluctuates about a constant value. This means that the components of Π_k and σ_k will grow large and the convergent part will be lost in rounding errors, since it is a difference of large values. These errors can be avoided if we choose an unconventional representation of Π_k and σ_k, which separates the convergent and divergent parts:

$$\Pi_k = \begin{pmatrix} A_k & y_k A_k \\ B_k & y_k B_k + t_k \end{pmatrix} , \tag{58}$$

$$\sigma_k = (0, 1)V_k + (1, y_k)U_k \quad . \tag{59}$$

As one eigenvalue of Π_k converges to zero, we will have $t_k \to 0$ while

$$A_k, B_k, U_k \to \infty \quad . \tag{60}$$

If we then choose

$$\begin{pmatrix} r \\ s \end{pmatrix} = \begin{pmatrix} y_k \\ -1 \end{pmatrix} , \tag{61}$$

we will have

$$\begin{pmatrix} a_k \\ b_k \end{pmatrix} = \frac{1}{y_k + \frac{\alpha_1}{\beta_1}} \begin{pmatrix} 0 \\ -t_k \end{pmatrix} \to 0 \tag{62}$$

and

$$v_k = -\frac{1}{y_k + \frac{\alpha_1}{\beta_1}} V_k \to H^{-1} x_1 \quad . \tag{63}$$

The relations (50,51,56,57) translate into the following relations for y_k, t_k, A_k, B_k, V_k and U_k:

$$\begin{aligned} y_1 &= 0 \quad , \\ t_1 &= 1 \quad , \\ A_1 &= 1 \quad , \\ B_1 &= 0 \quad , \\ V_1 &= -\frac{x_1}{\beta_1} \quad , \\ U_1 &= 0 \quad , \end{aligned} \tag{64}$$

$$\begin{pmatrix} A_{k+1} \\ \\ B_{k+1} \end{pmatrix} = \begin{pmatrix} -\frac{\alpha_{k+1}}{\beta_k} & 1 \\ \\ -\frac{\beta_{k+1}}{\beta_k} & 0 \end{pmatrix} \begin{pmatrix} A_k \\ \\ B_k \end{pmatrix} , \tag{65}$$

$$y_{k+1} = j_k + \frac{t_k}{A_{k+1}} , \tag{66}$$

$$t_{k+1} = -\frac{B_{k+1}}{A_{k+1}} t_k , \tag{67}$$

$$U_{k+1} = U_k + \frac{A_k}{\beta_k} x_{k+1} , \tag{68}$$

$$V_{k+1} = V_k - \frac{t_k}{A_{k+1}} U_{k+1} . \tag{69}$$

6. Application to Propagtor Calculations in Lattice QCD

Quark propagators in lattice QCD can be obtained from the inverse of the fermion matrix $M + 2ma$. As with the $<\bar{\psi}\psi>$ calculations, we can use the odd-even block structure of M to improve the calculation. We apply the Lanczos inversion algorithm to the matrix

$$H = i(M + 2ma) = \begin{pmatrix} 2mai & \hat{M} \\ \hat{M}^+ & 2mai \end{pmatrix} , \tag{70}$$

$$H^{-1} = \begin{pmatrix} -2mai[\hat{M}\hat{M}^+ + (2ma)^2]^{-1} & \hat{M}[\hat{M}\hat{M}^+ + (2ma)^2]^{-1} \\ \hat{M}^+[\hat{M}\hat{M}^+ + (2ma)^2]^{-1} & -2mai[\hat{M}\hat{M}^+ + (2ma)^2]^{-1} \end{pmatrix} . \tag{71}$$

The Lanczos equations are

$$\begin{aligned} Hx_1 &= 2maix_1 + \beta_1 x_2 , \\ Hx_i &= \beta_{i-1}x_{i-1} + 2maix_i + \beta_i x_{i+1} . \end{aligned} \tag{72}$$

The Lanczos vectors and the β's do not depend on 2 ma, so it is possible to obtain the inverse of a column at many different masses at the same time. This may be limited by storage space, in which case it is possible to calculate just the half of the column of the inverse which is in the diagonal block in equation (71). The other half can be reconstructed by multiplying by

$-\frac{1}{2mai}\hat{M}^+$ or $-\frac{1}{2mai}\hat{M}$ as appropriate.

In order to avoid a division by zero in the case $2ma=0$, we can use a slightly different representation of Π_k and σ_k than (58) and (59):

$$\Pi_{2k-1} = \begin{pmatrix} t_{2k-1} - (2ma)^2 y_{2k-1} B_{2k-1} & 2maiB_{2k-1} \\ 2maiy_{2k-1} A_{2k-1} & A_{2k-1} \end{pmatrix} ,$$

$$\Pi_{2k} = \begin{pmatrix} 2maiy_{2k} A_{2k} & A_{2k} \\ t_{2k} - (2ma)^2 y_{2k} B_{2k} & 2maiB_{2k} \end{pmatrix} , \qquad (73)$$

$$\sigma_k = (2maiV_k, 0) + (2maiy_k, 1)U_k \quad .$$

Then the factor 2mai divides out explicitly. This representation also has the advantage that the coefficients y_k, t_k, A_k and B_k are real.

We have investigated the convergence of t_k on 8^4 lattices using double precision arithmetic. We find that for $ma \geqslant 0.05$ at $\beta = 5.7$ convergence is similar to that of the conjugate gradient method. However, for zero mass or very small mass we find the convergence of the Lanczos algorithm is much better. For nonzero mass t_k converges steadily from the start of the calculation at a rate proportional to m. At zero mass the situation is very different. Then t_k shows no sign of convergence until we reach the point at which the first eigenvalues start to converge. From then on t_k drops rapidly and stops when convergence of the inverse is complete to within machine precision.

We are currently applying the algorithm to 16^4 lattices using single precision arithmetic. In order to check the accuracy we calculate:

$$r^2 = |Hz - x|^2 \quad , \qquad (74)$$

where z is the calculated value for $H^{-1}x$. We find

$$r^2 \lesssim 10^{-8} \quad .$$

7. Application to Fermion Updating

Probably the most difficult problem in lattice gauge theories is to include the effect of dynamical fermions on the gauge field configurations. This requires us to calculate the ratio of determinants of $M + 2ma$ between gauge configurations which differ only at one link. This must be done many times when the background gauge configuration is being generated. Because of the vast amount of computing that this takes, the quenched approximation has been used in the past in which the ratio of determinants has been taken as one.

This gives quite reasonable results but it is not ultimately satisfactory. There have been some useful attempts to overcome this using pseudo-fermions, microcanonical methods, etc., but once again the convergence at physically relevant masses is poor.

It is possible to use our inversion method in the updating problem as follows:

$$H = M + 2am \quad . \tag{75}$$

Changing one link changes H by ΔH:

$$R = \frac{\det\,(H + \Delta H)}{\det H} = \det\,(1 + H^{-1}\Delta H) \quad . \tag{76}$$

ΔH has only 18 nonzero elements, which lie within a 6×6 "block" at the intersection of the 6 rows and columns, corresponding to the 2 sites joined by the link which has been changed. It can be readily seen from this that only the corresponding 6×6 block of the inverse H^{-1} contributes in (76). Therefore, if we calculate 6 rows or columns of H^{-1}, we can update the link. Moreover, the same link can be updated a number of times without recalculating the inverse. There is some waste here in calculating a whole row when we only need 6 elements, but there is a way we can make use of it to improve the rate of convergence. Suppose for the moment that we have an approximation to the whole inverse:

$$Z = H^{-1} + \varepsilon \quad . \tag{77}$$

We can try to get an estimate of the error ε. Calculate

$$HZ = 1 + H\varepsilon \quad . \tag{78}$$

Then use the estimate of H^{-1} to calculate ε approximately:

$$Z(HZ - 1) = (H^{-1} + \varepsilon)H\varepsilon = \varepsilon + \varepsilon H\varepsilon \quad . \tag{79}$$

Then, provided Z was a good approximation, $\varepsilon H\varepsilon$ will be much smaller than ε and

$$Z - Z(HZ - 1) = H^{-1} - \varepsilon H\varepsilon \quad , \tag{80}$$

so we have a better approximation to the inverse from a relatively small amount of extra computation. If, now, we only have 6 columns of Z, it is still possible to do part of the calculation. First note that the rows of Z can be obtained trivially from the corresponding columns because of the hermitian–antihermitian structure in (71). Then it is possible to calculate the 6×6 elements of (80) that are needed for the updating.

We intend to apply this method using Lanczos to update on modest-sized lattices.

Acknowledgement. We would like to thank Rex Whitehead, Roger Haydock and Ian Duff for useful discussions. IMB and PEG would like to thank the DESY theory group for their hospitality during parts of the present work. We would also like to thank the SERC for use of the DAP facility at Queen Mary College and Prof. D. Parkinson and Kevin Smith for help in using it. PEG would like to thank the SERC for financial support and NB would like to thank the MESRS of Algeria for financial support. We would also like to thank J.P. Gilchrist and H. Schneider, our collaborators, for the lattice calculations.

References

1. J. Gilchrist, G. Schierholz, H. Schneider, M. Teper: Nucl. Phys. B **248**, 29 (1984)
2. K. Bowler, D. Chalmers, A. Kenway, R. Kenway, G. Pawley, D. Wallace: Edinburg preprint 84/295 (1984)
3. J. Cullum, R.A. Willoughby: In *Sparse Matrix Proceedings 1978*, ed. by I. Duff and G. Stewart (Siam Press, 1979);
   For a more recent variation of the method see:
   B.N. Parlett, J.K. Reed: IMA Journal of Numerical Analysis **1**, 135 (1981);
   See also: I.S. Duff: "A survey of sparse matrix software", Harwell report AERE-R 10512 (1982);
   B.N. Parlett: "The software scene in the extraction of eigenvalues from sparse matrices", Berkeley report PAM-132 (1983);
   R. Haydock: "Consequences of rounding error in the recursion and Lanczos methods", Cavendish preprint;
   B.N. Parlett: *The Symmetric Eigenvalue Problem* (Prentice-Hall, Englewood Cliffs 1980)
4. I. Barbour, P. Gibbs, J. Gilchrist, H. Schneider, G. Schierholz, M. Teper: Phys. Lett. **136**B, 80 (1984)
5. In preparation
6. This method has now been generalised and the eigenvalues of the nonsymmetric case found by a method analogous to the symmetric case. We thank Ian Duff for providing us with Cullum and Willoughby's routine for diagonalising the complex symmetric tridiagonal form

A Dedicated Lanczos Computer for Nuclear Structure Calculations

L.M. MacKenzie, D. Berry, A.M. MacLeod, and R.R. Whitehead

Department of Natural Philosophy, The University,
Glasgow G12 8QQ, United Kingdom

Using a combination of the occupation number representation and the Lanczos method, nuclear shell-model calculations can be cast in a form which is suitable for parallel computation. An attempt to design and construct the prototype of a suitable machine is described.

1 Introduction

This talk is about an attempt to design and build a dedicated computer for use in nuclear structure calculations. There is, of course, nothing new in the idea of dedicated computers - some people think that Stonehenge was one, and the Greeks certainly had them (the *antikythera* mechanism) as did the Arabs who invented the planispheric astrolabe. Mention of such devices is not completely irrelevant to the main topic of this conference; the original need for the development of rational approximation and continued fractions arose in connection with the gearing of planetaria and similar problems.

The thing that is relatively new, however, is the ease with which one can construct analogue computers out of digital bits and pieces. In effect, a modern analogue computer uses streams of digital numbers instead of electric currents or the rotation of a wheel as the analogue quantity.

The main requirement to be satisfied before a dedicated computer can be envisaged is that the calculations to be done must be cast in such a form that each step is as computationally well matched to the machinery as possible. Other speakers have already described how the matching or *mapping* is done in the case of lattice calculations using distributed array processors. A less obvious but more striking illustration is provided by the Fast Fourier Transform. In signal processing, where there is a natural desire and need to work in frequency space, progress was slow until the Fast Fourier Transform was introduced. Almost immediately thereafter, people were making dedicated on-line Fourier Transformers and the subject leapt ahead.

In the following sections we will discuss the nuclear shell model problem and describe the first attempt to build a computer whose structure matches as closely as possible the physics involved.

2 The Nuclear Shell Model

We use the expression "shell model" to refer to microscopic treatments of nuclear phenomena in which the elementary constituents are protons and neutrons. There are other kinds of nuclear models, but all of these must ultimately be referred back to the shell model just as the shell model must ultimately be referred back to the quark structure of the nucleons.

The essence of the shell model is that each nucleon is confined in a potential well produced by its interactions with all of the other nucleons. This well is often taken to be of the form of a three-dimensional harmonic oscillator as shown in Fig. 1. The ordering of and spacings between the various shells, os, op, 1sod, etc. account reasonably well for some of the gross properties of nuclei, and may be used as the foundation for configuration mixing studies.

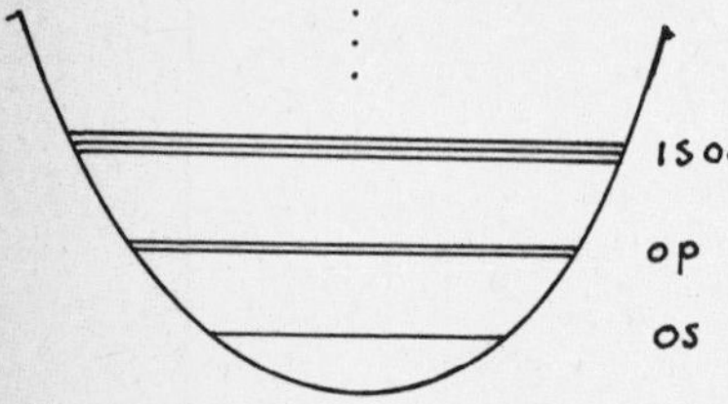

Figure 1 Schematic representation of the single-particle levels in a harmonic oscillator well

In the most usual approximation, only one major shell is actively involved in the configuration mixing. The computational problem is therefore to set up the Hamiltonian matrix evaluated between the states of the active configuration and then to diagonalise it. Both eigenvalues and eigenvectors are required, the latter to enable the calculation of transition rates and expectation value of various measurable quanities. Traditionally, that is since the mid 1930's, the basis states involved have been specified by means of group theory and the necessary matrix elements evaluated using Racah algebra and the formalism of fractional parentage. Such methods are very far from being matched in the sense described above.

The Lanczos method was first used in shell-model calculations in 1968 by SEBE and NACHAMKIN [1] and by WHITEHEAD [2]. Sebe and Nachamkin used it as a matrix diagonaliser but with the idea in mind that a well-chosen initial state would result in rapid convergence. Whitehead used it to calculate the tri-diagonal matrix directly from the two-body Hamiltonian without the intermediate step of constructing the full secular matrix. In both cases the basis states were specified group theoretically. A little later it was realised [3,4] that the standard formalism was an encumbrance, and that the full power of the Lanczos method could be brought to bear if the basis states and the Hamiltonian were specified in the occupation number representation:

$$|i\rangle = a^+_{i_1} a^+_{i_2} \dots a^+_{i_n} |0\rangle$$

$$\text{and } H = \sum_{\alpha\beta\gamma\delta} V_{\alpha\beta\gamma\delta} \; a^+_\alpha a^+_\beta a_\delta a_\gamma$$

where $|0\rangle$ represents the innert filled shells, the a's and a^+'s are fermion destruction and creation operators and the $V_{\alpha\beta\gamma\delta}$ are the two-body matrix elements that define H (there is, of course, also a one-body interaction, but it is computationally advantageous to combine it with the two-body part). The operation of multiplying a vector by H could now be performed using simple bit manipulations in the computer. For example, the state $|i\rangle$ can be represented by a string of 0's and 1's, the 1's representing the presence of the creation operators. When H operates on $|i\rangle$ each term in the sum results in a pair of 1's being removed and a new pair inserted.

The general organisation of such a calculation is illustrated in Fig. 2. The current vector is specified by a list of amplitudes for the basis states. Each basis state is operated on in turn by the Hamiltonian as outlined above and for each turn in H a new basis state results and the product of the initial amplitude A and the V involved is accumulated in the final amplitude vector B. The process as described is simply a matrix multiplication, but one in which the matrix is stored indirectly in a highly condensed form. There is certainly scope for parallel computation since a number of initial basis states could be handled simultaneously. Unlike some of the applications described

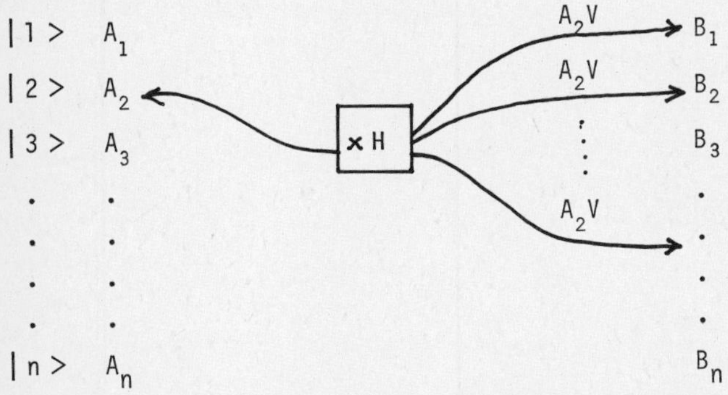

Figure 2

at this conference, though, it is the operation of multiplying a basis state by H rather than the arithmetic, the multiplication and accumulation of the A's and V's, that dominates the calculation. This is therefore not a suitable application for a single-instruction-multiple-data array processor.

3 The Prototype Machine

The advantages for shell-model work of a dedicated machine are:

(i) Low cost

(ii) Total access

(iii) Great computational power

The prototype machine to be described costs less than £10,000, will run reliably for long periods and has a performance comparable to that obtainable with an IBM 360/195. It is a quarter-scale version of the "production" machine, which will be capable of performing calculations that simply cannot be done on foreseeable commercial computers. It is nevertheless experimental in the sense that the final design is by no means fixed,and the prototype is intended as a testbed for future developments rather than as a finished machine.

The logical structure of the machine is shown in Fig. 3. The Matrix Format Generator performs the operations of creation and destruction and produces information about which A and which V (see Fig. 2) are to be multiplied and where the result is to be stored. This is passed to the Multiple Microprocessors Unit which performs the arithmetic, extracting the necessary data from and inserting the results in the Central Memory.

The Matrix Format Generator is shown schematically in Fig. 4. The Primary Generator constructs a basis state $|i>$ represented by a string of 32

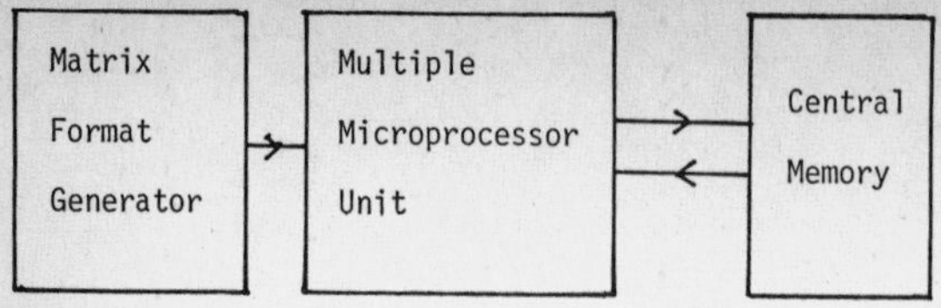

Figure 3 Logical structure of prototype machine

0's or 1's (the production version will have 128). This string is fed, in parallel, to the Secondary Generator where it acts as a "seed" stimulating the production of all the other basis states that have non-zero Hamiltonian matrix elements with the seed state. In the present version this is achieved by means of a system of self-addressing tables in which each 8-bit byte of the seed state is used as the address in a table at which a suitable target byte is to be found. This new byte is used in the same way until the original seed byte is again encountered signalling exhaustion of the possibilities.

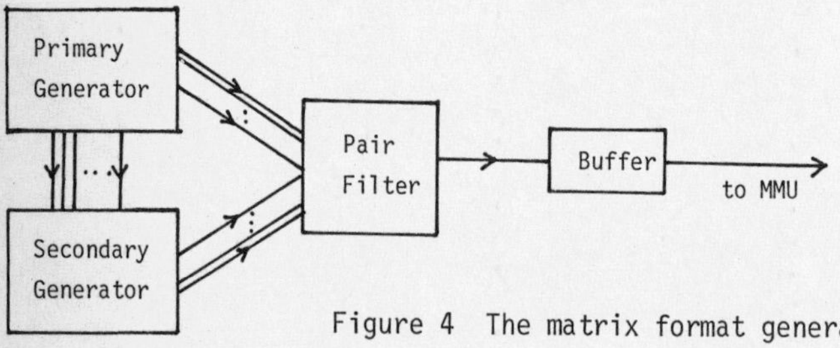

Figure 4 The matrix format generator

Owing to the conservation of additive quantum numbers, such as the third components of angular momentum and isospins, the Secondary Generator cannot be designed so as to produce *only* those basis states which have non-zero matrix elements with the seed state. It actually produces more states than it should. The function of the Pair Filter is to eliminate the redundant states and to extract the creation and destruction operators needed to convert the seed state into the target state. The indices of these operators specify which V is to be used later.

The Secondary Generator and Pair Filter are constructed from very fast Emitter-Coupled Logic components running at a clock rate of more than 100MHz. The output from the Pair Filter is buffered, to even out the rate of presentation to the Multiple Microprocessor Unit.

The design of the Matrix Format Generator was conditioned to a great extent by the relatively high cost of memory when the project began. The present design avoids the necessity to store the full list of basis states, which would have been very expensive in the projected 132-bit machine.

The output from the Matrix Format Generator, consisting of the index numbers of the initial and final basis states and the two-body matrix element indices, passes to the Multiple Microprocessor Unit. This consists of a set of identical microcomputers arranged so that whichever one is not busy accepts the next input and performs the necessary operations (see Fig. 5).

Figure 5

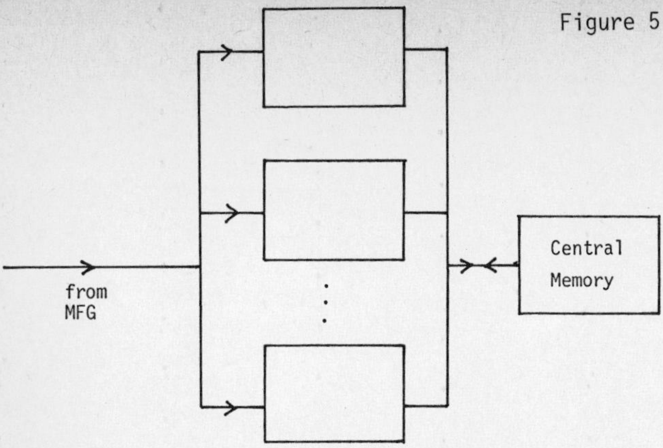

The tasks of extracting the relevant V and A, multiplying them together and storing the result cannot be accomplished by a single microprocessor without slowing down the Matrix Format Generator. The type of parallelism employed here is therefore one of overlapping operations in a series of asynchronous autonomous processors.

The prototype machine as described does not yet exploit all the possibilities for parallelism. For example, one could have two or more MFG's each working on different sections of the basis.

The machine was originally designed around 8-bit microprocessors for the sake of cheapness. It was, however, designed to be "upward compatible" with newer 16 and 32 bit microprocessors of the same (Motorola) family. These are very much faster and some have hardware floating point arithmetic. As a result of these advances, we now have designs for MMU modules, one or two of which will easily be able to keep up with the present MFG. This means that the MFG should now probably be redesigned. The cost of memory has also come down dramatically, and this may also have a bearing on future developments.

Acknowledgments

We are indebted to the Motorola Company for assistance in many aspects of this work. R.R.W. acknowledges the tenure of an SERC Senior Fellowship during the course of the work.

References

1. T. Sebe and J. Nachamkin Ann. Phys. (NY) 51 (1969) 100
2. R.R. Whitehead 1969 Unpublished report
3. R.R. Whitehead Nucl. Phys. A 182 (1972) 290
4. R.R. Whitehead, A. Watt, B.J. Cole and I. Morrison Adv. in Nucl. Phys. Vol. 9 Eds. Baranger and Voyt (Plenum Press, 1977)

Part VI

Conference Summary

Conference Summary

D.L. Weaire

Department of Pure and Applied Physics, Trinity College
Dublin, Ireland

E.P. O'Reilly

Department of Physics, University of Surrey, Guildford GU2 5XH, United Kingdom

1. The Scope of the Conference

Does the recursion method provide the best way of performing the kinds of calculations for which it was designed? How should it be implemented in practice? What rigorous mathematical results are available to guide us? What interesting applications are underway?

These were the main themes of the papers at this conference and the discussion which they provoked. We shall consider them in turn and try to give an impression of those intangible qualities which make conferences worthwhile - a sense of direction, a consensus regarding what is important and what is not, feelings of optimism or of disillusionment. In doing so, we shall reflect the contributed papers as well as the invited papers published in this volume. Our choice of highlights is a personal one, and we shall risk a few broad generalisations which may not cover all cases. We hope to be forgiven for this, in the interests of providing others with an appreciation of the outcome of the discussions.

2. Is it really the best way?

In describing the merits of the recursion method in their introductory papers, Heine [1] and Haydock [2] were largely preaching to the converted, but not entirely so.

Around the time when the recursion method was originally developed, at least two other methods were also suggested to calculate the local density of states. These were the method of moments and the equation-of-motion method. Until recently there have been comparatively few applications using either of these two methods. In the case of moments, this is because elementary applications of the method are numerically unstable, involving integrals of the form

$$\mu_k = \int_{-\infty}^{\infty} E^k \, n(E) \, dE,$$

while elementary applications of the equation-of-motion method require considerably more computing time than does the recursion method.

The paper on generalized moments by Gaspard [3] went a long way towards overcoming previous problems. The generalized-moments method is a cross between the moment method and the recursion method, retaining the property of linear dependence upon the density of states of the former, while being a stabler numerical technique.

A contributed paper by E Jurczek made explicit the link between the moments method and the recursion method. The recursion coefficients a_i and b_i can always be written in terms of Hankel determinants of the moments, but the calculation of these is numerically unstable. Jurczek showed how to choose a particular form of the generalized-moments method so that the recursion coefficients are related in a very simple way to the modified moments. The method is equivalent to generating a set of orthogonal but not orthonormal basis vectors using the recursion method.

Another form of real space method was described by Audit [4] using infinite generalised cyclic matrices. Applications of the method illustrated in a very accessible manner the physics of one-dimensional chains and of simple cubic crystals.

MacKinnon [5] advanced a case for the equation-of-motion method, which has generally been believed to be significantly less efficient than the recursion method. He argues, as we have done ourselves [6] in a different manner, that this is not necessarily the case. MacKinnon's conclusions were quite startling and remain to be analysed in more detail.

Weaire commented that the difficulty of the mathematical structure underlying the recursion method (the theory of continued fractions) was a disadvantage, in comparison with the equation-of-motion method, which uses Fourier transforms. Heine's reference to an esoteric result of Ramanujan, in his opening talk, exemplified beautifully the almost mystic quality of this subject for most of us. However, greater familiarity with continued fractions may make them more acceptable in future. There does not seem to be a suitable book which contains the essential mathematics relevant to our subject, without a good deal else.

Further comparisons with rival methods will undoubtedly be made in the years to come and will lead to some cross-fertilisation as well. For the time being the extreme simplicity of the recursion method earns it its pre-eminent place. But this simplicity is sometimes obscured by the various technical devices which have been added, to which we now turn.

How should it be implemented?

A major theme of this conference was undoubtedly the correct termination of a continued fraction. Much work has been devoted recently to understanding the relationship between the asymptotic form of the recursion coefficients and the density of states. Aspects represented in the conference were:

(i) analytic description of the effect of singularities and band gaps in the density of states upon the asymptotic forms of the recursion coefficients [7,8].

(ii) modified Gaussian quadrature techniques [9].

(iii) application of linear prediction methods to extrapolate recursion coefficients [10].

Together these provide a comprehensive description of how the coefficients behave asymptotically, and introduce new techniques for terminating calculations.

This remains a rather unsettled area. It is to be hoped that the analytical results will be extended to cover more cases - that of disordered Hamiltonians remains particularly obscure. The merits of practical procedures which have so far been proposed for termination are still regarded as debatable. Perhaps there will never be a perfect, all-purpose termination trick!

The simplicity of Allan's approach is certainly appealing, and he claimed that it could be readily adapted to suit special cases where there was available supplementary information which might be "built in" to the calculation.

Turchi [11] explained how perturbation theory might be used to advantage in comparing structural energies. Potentially, this offers advantages similar to those of pseudopotential perturbation theory in isolating the small, structure-dependent terms.

Jones [12] developed a generalised form of the recursion algorithm to calculate charge densities with a non-orthogonal basis and a block of starting states. This resulted in a block tridiagonal recursion scheme.

One problem clarified at the conference, at least for the authors of this summary, was that of the incorporation of broadening functions into the recursion method. This was explained by C Nex. The technique has been available for some years but has never been published, to our knowledge, so it may be worth outlining here.

Suppose we have calculated a continued function to N levels and wish to broaden the equivalent density of states. Using Gaussian quadrature we can calculate the poles x_i and weights ω_i corresponding to the given coefficients, so that the Green function G is approximated by

$$G(E) = \sum_{i=1}^{N} \frac{\omega_i}{E - x_i} , \tag{1}$$

and the density of states by N delta functions. The broadening function f(E) can also be approximated by N delta functions

$$f(E) = \sum_{j=1}^{N} f_j \ \delta(E-x_j) , \tag{2}$$

and (1) and (2) can be convoluted together to give

$$\begin{aligned} G^{br}(E) &= \int_{-\infty}^{\infty} G(E-E')\, f(E')\, dE' \\ &= \sum_{i=1}^{N} \sum_{j=1}^{N} \omega_i f_j \ \delta(E - x_i - x_j) . \end{aligned} \tag{3}$$

Inverse quadrature can then generate a set of N recursion coefficients equivalent to (3). These coefficients can be processed in the usual way to give the broadened density of states.

In the opinion of some this is not an important point, because it is a trivial exercise to directly convolute a calculated spectrum with any broadening function. This is true, but it seems "good style" to incorporate instrumental or other broadening functions at the outset, and not entirely without practical consequences. The mysteries of asymptotic behaviour are sometimes dispelled by this, since the effects of singularities are washed out.

3. Mathematical Results

We have already touched upon this aspect of the conference, since it is inevitably bound up with the consideration of termination procedures. However, it is worth noting that there was much appreciation of the contribution of Magnus, who has been delving into the mathematical depths of this subject for some time, and exhibited some of the nuggets of mathematical truth that he has found, in an invaluable table [7]. It will be interesting to see how long it takes to fill in and augment the table.

4. What Applications are Underway?

The conference did not contain any comprehensive review of the applications of the recursion method to date. Instead, it concentrated on current, and in some cases unorthodox, applications.

Off-diagonal Green functions were used in two papers. Jones [11] calculated charge densities using off-diagonal elements found directly in his block-triagonal scheme. Finnis [12] investigated force constants in a metal, calculating each off-diagonal element as the difference of two diagonal elements

$$G_{ij} = {}^{1}/{4}(G_{i+j,i+j} - G_{i-j,i-j})$$

Foulkes (in a contributed paper) showed an example from the electronic contribution to phonon frequencies where the recursion method was used to calculate the complicated k-space integral

$$I = \int_{\substack{E_k < E_F \\ E_{\underset{\sim}{k}+\underset{\sim}{g}} > E_F}} d^3k \, \frac{|M_{\underset{\sim}{k},\underset{\sim}{k}+\underset{\sim}{g}}|^2}{E_{\underset{\sim}{k}} - E_{\underset{\sim}{k}+\underset{\sim}{g}}} .$$

Maschke showed how the recursion method may be applied to self-consistent pseudopotential calculations, using it to calculate the lowest-lying eigenstates of a large Hamiltonian matrix. The method proceeded by choosing an arbitrary $|u_o\rangle$, then calculating $|u_1\rangle$ and the matrix

$$\begin{vmatrix} a_o & b_1 \\ b_1 & a_1 \end{vmatrix}$$

The lowest eigenstate of this matrix is chosen as the next $|u_o\rangle$ and the process can be iterated to converge rapidly. The eigenvectors calculated in one cycle can be used as starting states for the next cycle of a self-consistent calculation. This particular case may be better characterised as an application of the "Lanczos method" (see below), as are other calculations of electronic energy levels, which were described by Heine. Whatever it is called, the idea of combining two iterations in this way seems very attractive.

Annett presented model calculations to investigate Helium diffraction from corrugated metal surfaces.

Lambin described work using a second-neighbour linear chain Hamiltonian to examine, *inter alia*, the effects of internal singularities within a very elementary model.

O'Reilly used model calculations and approximate arguments to describe the linear scaling behaviour of the basis vector probability density $|u_n(\underset{\sim}{r})|^2$ for large n [13].

In a somewhat different category are the applications involving large sparse matrices in field theory and nuclear structure theory, described by Teper, Irving, Kenway and Whitehead [14-17]. In these, we are really dealing with the Lanczos Method, which uses the same algorithm to find individual eigenvalues as does the recursion method to generate smooth spectra.

The scale of the computational problems are considerably greater than those usually encountered in solid state physics with Kenway, for example, looking for eigenvalues in a 3145728 x 3145728 matrix! Two approaches in particular were being taken to extend the scope of the calculations: firstly, combining the Lanczos method with other, (variationally motivated) calculation schemes and secondly, using parallel processors and dedicated Lanczos computers.

The former approach is already widely used in solid state physics, for instance whenever a limited basis set is chosen or a cut-off placed on the range of interactions. The latter points to future directions for the recursion method. A dedicated computer, handling substantially smaller matrices and exploiting in full the parallel aspects of the recursion algorithm, would be of wide use and should be a goal for the future [17].

5. Future Directions

A number of outstanding problems were suggested during the conference. The application of the recursion method to disordered systems and its role in interpreting Anderson localization is not fully understood. It is not even agreed how the recursion coefficients behave asymptotically for a disordered Hamiltonian.

The relationship between the recursion method as $n \to \infty$ and the renormalization group has still to be interpreted, although some steps have been taken to do so.

Nor is it clear how to apply the recursion method to many-body problems where the density of states is in general unbounded, with $\langle u_0 | H^n | u_0 \rangle = \infty$ for some n, and where the basis states are not easily arranged into a countable set.

Roger Haydock lamented the absence of an analytic recursion model to describe the hydrogen atom. Analytic models are known for the free-electron gas and the simple harmonic oscillator and are useful for pedagogical purposes [18] and simple investigations [13]. The problems for the hydrogen atoms are due to the singularity of the Coulomb potential at r = 0.

All these problems have one feature in common: each involves an infinity of some kind. One of the major achievements of this conference has been the series of papers describing the asymptotic behaviour of the recursion coefficients for ordered Hamiltonians. It is not unrealistic to hope that future research will have similar success in describing the asymptotic behaviour of other aspects of the recursion method, and that greater understanding will continue to lead to better practical methods.

References

1. V Heine, this volume.
2. R Haydock, this volume.
3. J P Gaspard, this volume.
4. P Audit, this volume.
5. A MacKinnon, this volume.
6. D Weaire and E P O'Reilly, J Phys C, to be published (1984).
7. A Magnus, this volume.
8. F Ducastelle, this volume.
9. C Nex, this volume.
10. G Allan, this volume.
11. R Jones, this volume.
12. M W Finnis, this volume.
13. E P O'Reilly and D Weaire, J Phys A17, 2389 (1984).
14. M Teper, this volume.
15. A C Irving, this volume.
16. R D Kenway, this volume.
17. R R Whitehead, this volume.
18. R Haydock, Solid State Physics 35, 216 (1980).

Index of Contributors

P. Brüesch

Phonons: Theory and Experiments I

Lattice Dynamics and Models of Interatomic Forces

1982. 82 figures. XII, 261 pages. (Springer Series in Solid-State Sciences, Volume 34). ISBN 3-540-11306-1

Contents: Introduction. - Dynamics of the Linear Diatomic Chain. - Dynamics of Three-Dimensional Crystals. - Interatomic Forces and Phonon Dispersion Curves. - Anharmonicity. - Appendix. - General References. - References. - Subject Index.

H. Bilz, W. Kress

Phonon Dispersion Relations in Insulators

1979. 162 figures in 271 separate illustrations. VIII, 241 pages. (Springer Series in Solid-State Sciences, Volume 10)
ISBN 3-540-09399-0

Contents: Summary of Theory of Phonons: Introduction. Phonon Dispersion Relations and Phonon Models. - Phonon Atlas of Dispersion Curves and Densities of States: Rare-Gas Crystals. Alkali Halides (Rock Salt Structure). Metal Oxides (Rock Salt Structure). Transition Metal Compounds (Rock Salt Structure). Other Cubic Crystals (Rock Salt Structure). Cesium Chloride Structure Crystals. Diamond Structure Crystals. Zinc-Blende Structure Crystals. Wurtzite Structure Crystals. Fluorite Structure Crystals. Rutile Structure Crystals. ABO_3 and ABX_3 Crystals. Layered Structure Crystals. Other Low-Symmetry Crystals. Molecular Crystals. Mixed Crystals. Organic Crystals. - References. - Subject Index.

E. N. Economou

Green's Functions in Quantum Physics

2nd corrected and updated edition. 1983. 52 figures. XIV, 314 pages. (Springer Series in Solid-State Sciences, Volume 7). ISBN 3-540-12266-4

Contents: Green's Functions in Mathematical Physics: Time-Independent Green's Functions. Time-Dependent Green's Functions. - Green's Functions in One-Body Quantum Problem: Physical Significance of G. Application to the Free-Particle Case. Green's Functions and Perturbation Theory. Green's Functions for Tight-Binding Hamiltonians. Single Impurity Scattering. Two or More Impurities; Disordered Systems. - Green's Functions in Many-Body Systems: Definitions. Properties and Use of the Green's Functions. Calculational Methods for g. Applications. - Appendix A: Analytic Behavior of G(z) Near a Band Edge. - Appendix B: The Renormalized Perturbation Expansion (RPE). - Appendix C: Second Quantization. - References. - Subject Index.